FEYNMAN'S PATH INTEGRAL explained with basic Calculus

Published by SwaNi
Copyright © 2023 Swapnonil Banerjee
All rights reserved.

Preface

Feynman's Path Integral is a formulation of quantum mechanics that plays an important role in describing both non-relativistic (low speed) and relativistic (high speed) quantum phenomena, including quantum electrodynamics, and other quantum field theories. The Path Integral technique is usually considered an advanced topic reserved for graduate-level coursework. This book presents Path Integrals for non-relativistic particles from first principles with no more than basic Calculus. To avoid the mathematical rigor, complex Gaussian integrals are evaluated in a straightforward manner without detailed discussions on convergence (along the lines of the approach taken by Feynman and Hibbs in their book *Quantum Mechanics and Path Integrals*). I have also unapologetically appealed to common sense and analogies in my presentation. Feynman's Path Integral beautifully connects the classical with the quantum world. An object's classical trajectory in a force field plays an important role in the probabilistic world of quantum mechanics via a concept called the propagator as is discussed in this book.

Note that while this book demystifies Feynman's Path Integral by giving a gentle yet solid introduction to this rich concept, thorough discussions on the applications of Path Integral in various physical phenomena is beyond the scope of the book. Also, this book is not meant to be a substitute for a traditional textbook on quantum mechanics. This is meant as a gateway to Feynman's Path Integral and the beginning of your journey into this fascinating world. Empowered and inspired by what you learn here, you can study further on this remarkable topic from other more advanced literature. As a next step, I recommend the book *Quantum Mechanics and Path Integrals* by

Richard Feynman and Albert Hibbs, which acted as a main reference for my presentation.

Chapter 6 of this book teaches how to derive Schrodinger's equation, the key equation of non-relativistic quantum mechanics, using a propagator from first principles. There is an anecdote on that derivation that I had read in Feynman's Nobel lecture. Feynman learned of a paper by Paul Dirac at a beer party from a gentleman named Jehle. Pouring over the same together at a library the day next, to Jehle's utter astonishment, Feynman derived Schrodinger's equation in real-time based on an idea from that paper. The key concept in that derivation was the "propagator", which gave birth to the field of "Feynman's Path Integral." You will learn exactly how Feynman did it, in this book!

Prerequisites

A basic knowledge of Calculus is needed and should be adequate to follow this course. We will introduce Taylor expansion from the scratch, but knowing it beforehand will help (just a basic understanding will suffice). Some elementary knowledge of complex numbers is also necessary. Finally, since completing the square (a topic covered in school-level algebra) is widely used in the study of Path Integrals, a good grasp on completing the square will greatly facilitate understanding.

As for physics, school level classical mechanics, such as kinetic and potential energies, and Newton's law as a differential equation will be useful. No prior knowledge of quantum mechanics is required, although basic awareness of concepts, such as Planck's constant, de Broglie's wavelength, quantum wavefunction and its probabilistic connotation etc. will be useful.

In this book, "quantum mechanics" refers to "low speed", i.e., non-relativistic quantum particles. For the sake of accessibility, discussions are limited to non-relativistic particles and also any reference to Lagrangians is avoided. Variational Calculus has been used without explicitly referencing the Euler-Lagrange equation. All calculations are made from first principles. Every attempt has been made so that the core concepts of Path Integral can be followed with basic Calculus plus elementary knowledge of complex numbers.

Summary of Chapters

In Chapter 1, we will discuss all the major mathematical tools you will need for following this course. There will be many examples and exercises to help you.

In Chapter 2, we will discuss Variational Calculus, which will be useful in calculating Feynman's Path Integrals. We introduce the reader to the concept of action, and the Least Action Principle, based on concepts from Variational Calculus.

In Chapter 3, we will study Path Integrals from a mathematical point of view. You will see how a Path Integral is defined in terms of Gaussian integrals (part of the mathematical toolset covered in Chapter 1.)

In Chapter 4, we will explain how Path Integral shows up in quantum mechanics through the introduction of the concept of a Propagator. We will show how the classical trajectory of a particle is connected to the "quantum" propagator. We will calculate the propagator for a free particle (i.e., a particle without any forces acting on it) and use the propagator to derive de Broglie's wavelength and Einstein's energy-frequency relationship.

In Chapter 5, we will calculate the propagator for a particle when it has a potential energy in addition to having kinetic energy. We will consider potential energies that are linear and quadratic in the space variable respectively and evaluate the corresponding propagators. We will show how Variational Calculus becomes important in evaluating Path Integrals

4

indirectly. We will also show how to combine propagators, which will be greatly useful in chapter 6.

In Chapter 6, we will derive Schrodinger's equation from the propagator. The derivation will be based on the rule for combining propagators.

In Chapters 4, 5 and 6, our discussions of the Path Integral will revolve around three different physical systems: (1) Zero Potential Energy (2) Potential Energy varying linearly with space variable, and (3) Potential Energy varying quadratically with space variable. All our examples are limited to one spatial dimension.

In the final chapter (Chapter 7), we will show how the use of propagators in quantum mechanics shares formal similarity with combining probabilities in statistics. To emphasize the resemblance, we will discuss two different examples: one dealing with a cutting-edge technique used in biotechnology and the other, with a drunkard's walk. As we will see, the mathematics in both the cases are "formally similar" to the mathematics of propagators.

Swapnonil Banerjee, PhD
San Bruno, CA, USA.

Table of Contents

CHAPTER 1: MATHEMATICAL PREREQUISITES 2
    GAUSSIAN INTEGRALS 2
    SERIES EXPANSIONS 6
    COMPLEX NUMBERS 11
    BACK TO SERIES 12
    INTEGRALS WITH COMPLEX NUMBERS 14
    GAUSSIAN INTEGRALS WITH IMAGINARY NUMBERS 15
    ADDITIONAL EXERCISES 17

CHAPTER 2: VARIATIONAL CALCULUS 20
    STATIONARITY PRINCIPLE 24
    DERIVING THE EQUATION FOR THE SHORTEST PATH 26
    FURTHER EXAMPLES OF VARIATIONAL CALCULUS 30
    LEAST ACTION PRINCIPLE 37

CHAPTER 3: FEYNMAN'S PATH INTEGRAL: A MATHEMATICAL INTRODUCTION 44
    DISCRETIZING AN ACTION 44
    CALCULATING A SIMPLE PATH INTEGRAL 46
    ROLE OF THE CONSTANT 54
    INDIRECT METHOD FOR EVALUATING PATH INTEGRALS 56

CHAPTER 4: PATH INTEGRALS IN QUANTUM MECHANICS 64
    THE CONNECTION BETWEEN CLASSICAL AND QUANTUM 64
    DERIVATION OF DE BROGLIE'S AND EINSTEIN'S RELATION FROM FREE PARTICLE PROPAGATOR . 74

CHAPTER 5: PATH INTEGRALS IN THE PRESENCE OF POTENTIAL ENERGIES 78
    POTENTIAL ENERGIES IN THE MICROSCOPIC WORLD 78
    PATH INTEGRALS WITH THE POTENTIAL ENERGY TERM IN THE ACTION 80
    MULTIPLICATIVE LAW OF PROPAGATORS 84
    EXAMPLES OF CALCULATING PATH INTEGRALS WITH POTENTIAL ENERGY 89

CHAPTER 6: DERIVING SCHRODINGER'S EQUATION USING THE PROPAGATOR .. 100
    EXAMPLE-1: DERIVING SCHRODINGER'S EQUATION FOR A FREE PARTICLE 103
    EXAMPLE-2: DERIVING SCHRODINGER'S EQUATION FOR A PARTICLE HAVING A POTENTIAL ENERGY: 106
        Case I: The Potential Energy V(x) Is Linear in the Space Variable 107
        Case II: The Potential Energy V(x) Is Quadratic in the Space Variable 110
        Case-III: Deriving Schrodinger's Equation for a Particle with a General Potential Energy V(x) 113

CHAPTER 7 CLASSICAL PROBABILITIES AND QUANTUM PROBABILITY AMPLITUDES ... 118

- INTRODUCTION ... 118
- EXAMPLE 1: DIGITAL PCR, CALCULATING CONCENTRATION FROM THE ODDS OF NEGATIVES .. 120
- POISSON PLUS—A DIGITAL PCR ALGORITHM ... 123
- EXAMPLE 2: RANDOM WALK ... 128

APPENDIX 1 ... 132

- FUNCTIONALS ... 132

APPENDIX 2 ... 134

- POTENTIAL ENERGIES ... 134

APPENDIX 3 ... 136

- HEISENBERG'S UNCERTAINTY PRINCIPLE: AN APPLICATION OF THE MULTIPLICATIVE LAW OF PROPAGATORS (ADVANCED TOPIC) ... 136

APPENDIX 4 ... 142

- THE WIDE-RANGING APPLICABILITY OF FREE PARTICLE PROPAGATORS (ADVANCED TOPIC) 142

APPENDIX 5 ... 146

- FUNCTIONAL DERIVATIVES ... 146

APPENDIX 6 ... 152

- STATISTICS OF A RANDOM WALK ... 152

APPENDIX 7 ... 156

- SOLUTIONS ... 156

INDEX ... 170

Chapter 1: Mathematical Prerequisites

Gaussian Integrals

Let us consider one of the most important mathematical functions encountered in physics and statistics: the Gaussian function. It is given by $f(x) = e^{-x^2}$ and shown in Figure 1. The area under the curve $f(x) = e^{-x^2}$ from $-\infty$ to $+\infty$ is $\sqrt{\pi}$ as shown in Eq. (1):

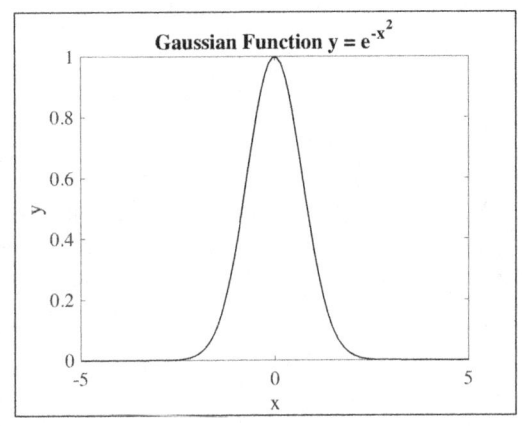

Figure 1

$$\int_{-\infty}^{+\infty} e^{-x^2} dx = \sqrt{\pi} \quad \text{... Eq. (1)}.$$

We urge you to memorize this result in the same spirit as you remember, say, $\sin\left(\frac{\pi}{4}\right) = \frac{1}{\sqrt{2}}$. Several applications of Eq. (1) are given as follows:

Examples

Example 1. Find $\int_{-\infty}^{+\infty} e^{-4x^2} dx$ using Eq. (1).

<u>Soln.</u> Since, $4x^2 = (2x)^2$, make the following change of variable: $u \equiv 2x$, implying $du = 2dx \Rightarrow dx = \frac{du}{2}$.

Hence, $\int_{-\infty}^{+\infty} e^{-4x^2} dx = \int_{-\infty}^{+\infty} e^{-4x^2} dx = \frac{1}{2}\int_{-\infty}^{+\infty} e^{-u^2} du = \frac{1}{2}\sqrt{\pi}$. In the last step, we used Eq. (1), i.e., $\int_{-\infty}^{+\infty} e^{-u^2} du = \sqrt{\pi}$.

Example 2. Prove $\int_{-\infty}^{+\infty} e^{-\frac{x^2}{2\sigma^2}} dx = \sigma\sqrt{2\pi}$.

2

Feynman's Path Integral explained with basic Calculus

<u>Soln.</u> Make the following change of variable: $u = \frac{x}{\sqrt{2}\sigma}$, such that $u^2 = \frac{x^2}{2\sigma^2}$.

From $u = \frac{x}{\sqrt{2}\sigma}$, you have $du = \frac{dx}{\sqrt{2}\sigma} \Rightarrow dx = \sqrt{2}\sigma du$. Hence,

$$\int_{-\infty}^{+\infty} e^{-\frac{x^2}{2\sigma^2}} dx = \sigma\sqrt{2} \int_{-\infty}^{+\infty} e^{-u^2} du$$

$$= \sigma\sqrt{2}\sqrt{\pi} \quad \text{(We used Eq. (1): } \int_{-\infty}^{+\infty} e^{-u^2} du = \sqrt{\pi}\text{)}$$

$$= \sigma\sqrt{2\pi} \quad \text{(Proved.)}$$

How does $f(x) = e^{-\frac{x^2}{2\sigma^2}}$ look? It is similar to e^{-x^2}, but its width depends on the parameter σ. For example, if σ is large, $e^{-\frac{x^2}{2\sigma^2}}$ looks wide, and if σ is small, $e^{-\frac{x^2}{2\sigma^2}}$ looks narrow.

What about $\int_{-\infty}^{+\infty} e^{-\frac{(x-a)^2}{2\sigma^2}} dx$? The integrand is the function $e^{-\frac{x^2}{2\sigma^2}}$ shifted to the right by a units. If we make the following change of variable: $y = x - a$, then $y = -\infty$ when $x = -\infty$, and $y = +\infty$ when $x = +\infty$. Hence,

$$\int_{-\infty}^{+\infty} e^{-\frac{(x-a)^2}{2\sigma^2}} dx = \int_{-\infty}^{+\infty} e^{-\frac{y^2}{2\sigma^2}} dy,$$ which we evaluated in the previous problem (Example 2) to be $\sigma\sqrt{2\pi}$. Therefore, we have,

$$\int_{-\infty}^{+\infty} e^{-\frac{(x-a)^2}{2\sigma^2}} dx = \sigma\sqrt{2\pi} \quad \dots\dots\dots\dots\dots\dots\dots\dots \text{Eq. (2).}$$

If we divide both sides of Eq. (2) by $\sigma\sqrt{2\pi}$, we get:

$$\int_{-\infty}^{+\infty} \frac{1}{\sigma\sqrt{2\pi}} e^{-\frac{(x-a)^2}{2\sigma^2}} dx = 1.$$

Note how the integral of $\frac{1}{\sigma\sqrt{2\pi}} e^{-\frac{(x-a)^2}{2\sigma^2}}$ is equal to 1. It is because of this reason that the Gaussian function $e^{-\frac{(x-a)^2}{2\sigma^2}}$ is said to be "normalized" when divided by the factor $\sigma\sqrt{2\pi}$, and the normalized Gaussian function is given by:

Chapter 1: Mathematical Prerequisites

$$\frac{1}{\sigma\sqrt{2\pi}} e^{-\frac{(x-a)^2}{2\sigma^2}} dx \quad \ldots \ldots \ldots \ldots \ldots \ldots \ldots \ldots \ldots \ldots \ldots \ldots \ldots \ldots \ldots \text{Eq. (3)}.$$

Eq. (3) will play a vital role in our discussion of Path Integrals.

RECAP: Normalized Gaussian: $f(x) = \dfrac{1}{\sigma\sqrt{2\pi}} e^{-\frac{(x-a)^2}{2\sigma^2}}$

Why is it called normalized?

Because the integral of the function from $x = -\infty$ to $x = +\infty$ is 1!

$$\int_{-\infty}^{+\infty} \frac{1}{\sigma\sqrt{2\pi}} e^{-\frac{(x-a)^2}{2\sigma^2}} dx = 1$$

Example 3. As yet another application of Eq. (1), let us evaluate $\int_{-\infty}^{+\infty} e^{-2x^2+4x} dx$.

Soln. In order to use Eq. (1) for this problem, you will first need to "complete the square" for $-2x^2 + 4x$.

$$-2x^2 + 4x$$
$$= -2(x^2 - 2x)$$
$$= -2(x^2 - 2 \cdot x \cdot 1 + 1^2 - 1)$$
$$= -2((x-1)^2 - 1) \quad [\text{We used } (a+b)^2 = a^2 + 2ab + b^2]$$
$$= -2(x-1)^2 + 2 \quad \ldots \ldots \ldots \ldots \ldots \ldots \ldots \ldots \ldots \ldots \ldots \ldots \text{Eq. (4)}.$$

Hence, using Eq. (4) in $\int_{-\infty}^{+\infty} e^{-2x^2+4x} dx$, we get:

$$\int_{-\infty}^{+\infty} e^{-2x^2+4x} dx$$
$$= \int_{-\infty}^{+\infty} e^{-2(x-1)^2 + 2} dx$$
$$= \int_{-\infty}^{+\infty} e^{-2(x-1)^2} \cdot e^2 dx$$
$$= e^2 \int_{-\infty}^{+\infty} e^{-2(x-1)^2} dx \quad \ldots \ldots \ldots \ldots \ldots \ldots \ldots \ldots \ldots \text{Eq. (5)}.$$

Next, substituting $y = \sqrt{2}(x-1)$, and $dy = \sqrt{2} dx$ in Eq. (5), we get:

4

Feynman's Path Integral explained with basic Calculus

$\int_{-\infty}^{+\infty} e^{-2x^2+4x} dx = \frac{e^2}{\sqrt{2}} \int_{-\infty}^{+\infty} e^{-y^2} dy = \frac{e^2}{\sqrt{2}} \sqrt{\pi}$. [We used $\int_{-\infty}^{+\infty} e^{-y^2} dy = \sqrt{\pi}$, as obtained in Eq. (1).]

A Peculiar Feature of Gaussian Integrals

In the last example, you got $\int_{-\infty}^{+\infty} e^{-2x^2+4x} dx = e^2 \sqrt{\frac{\pi}{2}}$. You have e^{-2x^2+4x} as the integrand on the left of the equation and e^2 times a constant, on the right. Now, "2", the argument of the exponential e^2 on the right is the maximum of $-2x^2 + 4x$, as can be seen by completing the square: $-2x^2 + 4x = -2(x-1)^2 + 2$, in which $(x-1)^2$, being a squared term, is either positive or zero. Since, the maximum of $-2(x-1)^2$ is zero, $-2x^2 + 4x$ maxes at 2.

Generally speaking, for any integral of the form $\int_{-\infty}^{+\infty} e^{-Ax^2+Bx} dx$ $(A > 0)$, you can (partially) evaluate the integral by finding the maximum value of $-Ax^2 + Bx$ and using the maximum in the exponent of e. Then $\int_{-\infty}^{+\infty} e^{-Ax^2+Bx} dx = e^{(maximum \, of \, -Ax^2+Bx)}(constant)$.

To solidify this concept, let us see another example. We will determine $\int_{-\infty}^{+\infty} e^{-2x^2+6x} dx$ up to a constant. Completing the square, $-2x^2 + 6x = -2(x^2 - 3x) = -2\left(x^2 - 2.x.\frac{3}{2} + \left(\frac{3}{2}\right)^2 - \frac{9}{4}\right) = -2\left(x - \frac{3}{2}\right)^2 + \frac{9}{2}$, which is maximum when the square term is zero. Hence, the maximum value of $-2x^2 + 6x$ is $\frac{9}{2} = 4.5$, using which in the exponent of e, we get:

$\int_{-\infty}^{+\infty} e^{-2x^2+6x} dx = $ (a constant)* $e^{4.5}$.

Exercise 1. Evaluate $\int_{-\infty}^{+\infty} e^{-2x^2+6x} dx$ by directly completing the square, then using Eq. (1), i.e., $\int_{-\infty}^{+\infty} e^{-x^2} dx = \sqrt{\pi}$. Does the integral come out as (a constant)* $e^{4.5}$? Solution available in Appendix 7.

Chapter 1: Mathematical Prerequisites

A General Result

The following can be proved by completing the square that for $A > 0$,

$$\int_{-\infty}^{+\infty} e^{-Ax^2+Bx} dx = \sqrt{\frac{\pi}{A}} e^{\frac{B^2}{4A}} \quad \text{................Eq. (6).}$$

Exercise 2. Prove Eq. (6). Solution available in Appendix 7.

A word about the convergence of Eq. (6): A must be positive for the integral to converge, which you can argue as follows. The integrand e^{-Ax^2+Bx} can be written as $e^{-Ax^2} e^{Bx}$. When $A > 0$, e^{-Ax^2} is much smaller than the other factor e^{Bx} for large $|x|$ ($x^2 \gg |x|$). As a result, e^{-Ax^2+Bx} goes to zero as $|x|$ becomes larger and larger. However, if $A < 0$, e^{-Ax^2} will grow without bound as x goes to infinity, thereby making the integral in Eq. (6) divergent.

RECAP: An Important Integral: $\int_{-\infty}^{+\infty} e^{-Ax^2+Bx} dx = \sqrt{\frac{\pi}{A}} e^{\frac{B^2}{4A}}$

Example 4. We will calculate $\int_{-\infty}^{+\infty} e^{-2x^2+4x} dx$ using Eq. (6).

<u>Soln.</u> We have, $-2x^2 + 4x \equiv -Ax^2 + Bx$, which implies $A = 2$ and $B = 4$. Hence, using Eq. (6), $\int_{-\infty}^{+\infty} e^{-2x^2+4x} dx = \sqrt{\frac{\pi}{2}} e^{\frac{4^2}{(4)(2)}} = \sqrt{\frac{\pi}{2}} e^2$.

Series Expansions

Binomial Expansion

How do you find $(1 + x)^2$? You simply multiply $(1 + x)$ by $(1 + x)$ to obtain $(1 + x)(1 + x) = 1 + 2x + x^2$. What about $(1 + x)^3$? You can check by multiplying $(1 + x)$ to $(1 + x)^2$ that $(1 + x)^3 = 1 + 3x + 3x^2 + x^3$. The general formula for $(1 + x)^n$ is:

Feynman's Path Integral explained with basic Calculus

$$(1+x)^n = 1 + nx + \frac{n(n-1)}{2!}x^2 + \frac{n(n-1)(n-2)}{3!}x^3 + \frac{n(n-1)(n-2)(n-3)}{4!}x^4 + \cdots$$

$$\cdots\cdots\cdots\cdots\cdots\cdots\cdots\cdots\cdots\cdots\cdots\cdots\cdots \text{Eq. (7)}.$$

Let us check the validity of the above formula for $n = 3$. The third term on the right side has the factor $\frac{n(n-1)}{2!}$, which, when evaluated at $n = 3$, gives $\frac{3(3-1)}{2!} = 3$. For the next term, $\frac{n(n-1)(n-2)}{3!} = \frac{3(3-1)(3-2)}{3 \cdot 2} = 1$. Then, setting $n = 3$ in $\frac{n(n-1)(n-2)(n-3)}{4!}x^4$, we get zero; all the subsequent terms of the series in Eq. (7) are zero, due to the presence of the factor $(n-3)$ in all of them. Hence, the series of Eq. (7) is terminated at the x^3 term producing: $(1+x)^3 = 1 + 3x + 3x^2 + x^3$, the same as what you got by direct multiplication.

Now, the cool thing about Eq. (7) is that it is correct even when n is not an integer. For example, you can set $n = \frac{1}{2}$ in Eq. (7). Then on the left side, you will have $(1+x)^{\frac{1}{2}}$. And on the right, you will have the following series for $(1+x)^{\frac{1}{2}}$:

$$(1+x)^{\frac{1}{2}} = 1 + \frac{1}{2}x + \frac{\frac{1}{2}(\frac{1}{2}-1)}{2!}x^2 + \frac{\frac{1}{2}(\frac{1}{2}-1)(\frac{1}{2}-2)}{3!}x^3 + \cdots$$

$$= 1 + \frac{1}{2}x - \frac{1}{8}x^2 + \frac{1}{16}x^3 + \cdots \cdots\cdots\cdots\cdots \text{Eq. (8)}.$$

Note that the above series never ends and hence is an infinite series. The series converges to a finite value when the absolute value of x is less than 1. You can appreciate that by noticing that the higher order terms (corresponding to larger powers of x) of the series get smaller and smaller when x is a fraction.

Eq. (7) is sometimes referred as the Binomial expansion. The name "Binomial" refers to the "two" terms in $1 + x$.

Chapter 1: Mathematical Prerequisites

Exercise 3. Taking $x = 0.4$, calculate an approximate value for $\sqrt{1.4}$ using Eq. (7). (You can keep terms up to the third power of x.) Then, calculate $\sqrt{1.4}$ with your calculator and verify that the two results are close.

Exercise 4. Find the numerical coefficient for the x^5 term of $(1+x)^{\frac{1}{2}}$, using Eq. (7).

Taylor Expansion

Next, we show you how the series expansion of $(1+x)^n$ (given by Eq. (7)) is obtained with a technique called the Taylor expansion.

Let $f(x) = (1+x)^n$. Differentiating $f(x)$ multiple times, we get:

$\frac{df}{dx} = n(1+x)^{n-1}$, $\frac{d^2f}{dx^2} = n(n-1)(1+x)^{n-2}$, $\frac{d^3f}{dx^3} = n(n-1)(n-2)(1+x)^{n-3}$,....etc.

Evaluating $f(x)$ and its derivatives at $x = 0$, we get:

$f(x)|_{x=0} = 1$, $\frac{df}{dx}\Big|_{x=0} = n$, $\frac{d^2f}{dx^2}\Big|_{x=0} = n(n-1)$, $\frac{d^3f}{dx^3}\Big|_{x=0} = n(n-1)(n-2)$,......etc.

Note that $f(x)$ and its derivatives evaluated above show up in the Binomial expansion of $f(x) = (1+x)^n$ as given by Eq. (7). For example, the first term on the right of Eq. (7) is $f(x)|_{x=0}$. The second term is nx, which can be written as $\frac{df}{dx}\Big|_{x=0} x$, the third term, $\frac{n(n-1)}{2!}x^2$, can be written as $\frac{1}{2!}\frac{d^2f}{dx^2}\Big|_{x=0} x^2$, and so on. Hence, the expansion of $f(x) = (1+x)^n$ in Eq. (7) can be expressed as:

$$f(x) = f(x)|_{x=0} + \frac{df}{dx}\Big|_{x=0} x + \frac{1}{2!}\frac{d^2f}{dx^2}\Big|_{x=0} x^2 + \frac{1}{3!}\frac{d^3f}{dx^3}\Big|_{x=0} x^3 + \cdots$$

.... Eq. (9).

Although we checked Eq. (9) for $f(x) = (1+x)^n$, Eq. (9) is valid for any smooth function $f(x)$ expanded about $x = 0$. Such an expansion goes by the

Feynman's Path Integral explained with basic Calculus

name of Taylor's expansion. As you can see, you can call the expansion of $(1+x)^n$ as either Binomial or Taylor.

As another example of Taylor expansion, consider $f(x) = e^x$, which we want to expand about $x = 0$. We have, $f(x)|_{x=0} = 1$, $\frac{df}{dx}\big|_{x=0} = e^x|_{x=0} = 1$.

Similarly, all the other derivatives, such as $\frac{d^2f}{dx^2}$ and $\frac{d^3f}{dx^3}$ are 1 at $x = 0$. Hence, using the Taylor expansion about $x = 0$, as given by Eq. (9), we have:

$$e^x = 1 + x + \frac{x^2}{2!} + \frac{x^3}{3!} + \cdots \qquad \text{Eq. (10)}.$$

Exercise 5. Expand $f(x) = (1+x)^{-1}$ about $x = 0$ using Eq. (9). If you think your Taylor expansion looks like an infinite GP series, you are right. What is the common ratio of the GP series you just found? What is the range of values of x for which the series converges? [Hint: the magnitude of the common ratio of an infinite GP series should be less than 1 for the series to converge.]

In the Taylor expansion of $f(x)$, given by Eq. (9), there is nothing special about $x = 0$. We could have Taylor-expanded a smooth function $f(x)$ about any other value of x, say $x = a$, by modifying Eq. (9) as follows.

$$f(x) = f(x)|_{x=a} + \frac{df}{dx}\bigg|_{x=a}(x-a) + \frac{1}{2!}\frac{d^2f}{dx^2}\bigg|_{x=a}(x-a)^2 +$$

$$\frac{1}{3!}\frac{d^3f}{dx^3}\bigg|_{x=a}(x-a)^3 + \cdots \qquad \text{Eq. (11)}.$$

What is the range of values of x for which such a series valid? That depends on the function you are Taylor-expanding. For example, the series expansion of $f(x) = \frac{1}{1-x}$ about $x = 0$ is valid for $0 < |x| < 1$. On the other hand, for $f(x) = e^x$, the expansion given by Eq. (10) is valid for all values of x. The

9

Chapter 1: Mathematical Prerequisites

Taylor-series of Eq. (11) is usually useful for practical applications when x is close to a, the point the function is expanded about.

Example 5. We will calculate $e^{2.1}$ using Eq. (11). Let $f(x) = e^x$. Hence, $e^{2.1} = f(x = 2.1)$. We will expand $f(x)$ about $x = 2$, i.e., for us, a, per Eq. (11), is 2.

<u>Soln.</u> Now, $f(x) = e^x, \frac{df}{dx} = \frac{d^2f}{dx^2} = \cdots = e^x$. Evaluating the function and its derivatives at $x = 2$, we obtain from Eq. (11):

$$f(x) = e^2 + (x-2)e^2 + \frac{(x-2)^2}{2!}e^2 + \cdots$$

Next setting $x = 2.1$ in the above, we get:

$$f(2.1) = e^2 + (2.1-2)e^2 + \frac{(2.1-2)^2}{2!}e^2 + \cdots$$

$$= e^2 + (.1)e^2 + \frac{(.1)^2}{2!}e^2 + \cdots$$

$$= e^2\left(1 + .1 + \frac{(.1)^2}{2!} + \cdots\right) \approx 8.165, \text{ which is close to } e^{2.1} \approx 8.166.$$

There is nothing special about expanding e^x about 2 to approximately evaluate $e^{2.1}$. We could have expanded e^x about any other number, e.g., 1.5. But then, we would have needed a larger number of terms to get the same level of accuracy. In Eq. (11), the closer x is to a, the smaller is the number of terms required for the same level of accuracy.

RECAP: <u>Taylor Expansion of a Function $f(x)$ about $x = a$:</u>

$$f(x) = f(x)|_{x=a} + \frac{df}{dx}\bigg|_{x=a}(x-a) + \frac{1}{2!}\frac{d^2f}{dx^2}\bigg|_{x=a}(x-a)^2$$
$$+ \frac{1}{3!}\frac{d^3f}{dx^3}\bigg|_{x=a}(x-a)^3 + \cdots$$

Feynman's Path Integral explained with basic Calculus

Complex Numbers

We will now briefly discuss Complex numbers, which we will need in Feynman's Path Integrals. The imaginary number i is defined as $i \equiv \sqrt{-1}$, squaring which, you get $i^2 = -1$. A complex number has both real and imaginary parts. For example, the complex number $2 + 3i$ has 2 for the real part and $3i$ for the imaginary part. If you have an equation $a + bi = 2 + 3i$, then you can equate the real part a on the left side to the real part 2 on the right and the imaginary part bi to $3i$. So, $a = 2$ and $bi = 3i$. Cancelling i from both sides of the last equation, we get, $b = 3$.

Matching real to real and the imaginary to imaginary in an equation is a powerful technique. We will demonstrate its use through the following example. We know that i is the square root of -1, but what is the square root of i?

Let $\sqrt{i} = a + bi$, where a and b are real numbers. We will determine a and b as follows.

Squaring both sides of the equation $\sqrt{i} = a + bi$, we get:

$i = (a + bi)^2$

$\Rightarrow i = (a^2 - b^2) + 2abi$ [We expanded $(a + bi)^2$ using

$(x + y)^2 = x^2 + 2xy + y^2$, and used $i^2 = -1$.]

Now, there is no real number on the left side of the equation in the last line, whereas on the right, you have the real number $(a^2 - b^2)$. So, equating real to real, we get, $(a^2 - b^2) = 0$. Next, equating the imaginary number i on the left to the imaginary number $2abi$ on the right, we get $i = 2abi$. Cancelling i from both sides, $2ab = 1$.

Summarizing, we have two equations: $(a^2 - b^2) = 0$ and $2ab = 1$. We will now solve these to obtain a and b. From $(a^2 - b^2) = 0$, we get, $a = \pm b$. Using $a = -b$ in the second equation, viz., $2ab = 1$, we get $-2a^2 = 1$, which

Chapter 1: Mathematical Prerequisites

does not have a solution for any real a. So, we try $a = b$, using which in $2ab = 1$, we get: $a^2 = \frac{1}{2} \Rightarrow a = \pm\frac{1}{\sqrt{2}}$. b will also be $\pm\frac{1}{\sqrt{2}}$, since $b = a$.

Hence, we have the following solutions:

1) $a = +\frac{1}{\sqrt{2}}, b = +\frac{1}{\sqrt{2}}$, 2) $a = -\frac{1}{\sqrt{2}}, b = -\frac{1}{\sqrt{2}}$. In other words, there are two possible values of $\sqrt{i}$:

1. $\frac{1}{\sqrt{2}} + i\frac{1}{\sqrt{2}}$
2. $-\frac{1}{\sqrt{2}} - i\frac{1}{\sqrt{2}}$

Exercise 6. Check by squaring 1 and 2 in the above that you get i in each case.

Back to Series

We return to the Taylor expansion. You have already seen how to Taylor expand e^x. In this section, we will introduce you to the Taylor expansion of e^{ix}, a function that is ubiquitous in Path Integrals. For calculating the Taylor expansion of e^{ix}, you will need Taylor expansions of $\sin x$ and $\cos x$ about $x = 0$, which we calculate first.

Taylor Expansion of $\sin x$ about $x = 0$

We will use $f(x) = \sin x$, and $a = 0$ in Eq. (11).

$f(x) = \sin x \Rightarrow \frac{df}{dx} = \cos x, \frac{d^2f}{dx^2} = -\sin x, \frac{d^3f}{dx^3} = -\cos x, \ldots$ etc.

Hence, $f(x)|_{x=0} = \sin x|_{x=0} = 0$, $\frac{df}{dx}|_{x=0} = \cos x|_{x=0} = 1, \ldots$etc.

Using $f(x)$ and its derivatives at $a = 0$, as calculated above, we obtain from Eq. (11):

$$\sin x = x - \frac{x^3}{3!} + \cdots \quad \text{... Eq. (12).}$$

Feynman's Path Integral explained with basic Calculus

Exercise 7. Obtain the first few terms of the Taylor expansion of $f(x) = \cos x$ using Eq. (11). You should get:

$$\cos x = 1 - \frac{x^2}{2!} + \frac{x^4}{4!} - \cdots \qquad \text{Eq. (13).}$$

Taylor Expansion of $e^{i\theta}$

In Eq. (10), we derived the Taylor expansion of e^x about $x = 0$, which we rewrite in the following:

$$e^x = 1 + x + \frac{x^2}{2!} + \frac{x^3}{3!} + \frac{x^4}{4!} \cdots \qquad \text{Eq. (10).}$$

If you set $x = i\theta$ on the left of Eq. (10), you have $e^{i\theta}$. What would you get on the right? To answer that, you need the following quantities.

$x = i\theta$

$x^2 = (i\theta)^2 = -\theta^2$ [We used $i^2 = -1$]

$x^3 = (i\theta)^3 = -i\,\theta^3$ [$i^3 = -i$]

$x^4 = (i\theta)^4 = \theta^4$

.... etc.

Using x, x^2, x^3...etc., as calculated above, in Eq. (10), we get:

$$e^{i\theta} = 1 + i\theta + \frac{(i\theta)^2}{2!} + \frac{(i\theta)^3}{3!} + \frac{(i\theta)^4}{4!} + \cdots$$

$$= 1 + i\theta - \frac{\theta^2}{2!} - i\frac{\theta^3}{3!} + \frac{\theta^4}{4!} + \cdots$$

$$= \left(1 - \frac{\theta^2}{2!} + \frac{\theta^4}{4!} + \cdots\right) + i\left(\theta - \frac{\theta^3}{3!} + \cdots\right) \text{ [Grouping the real and the imaginary terms separately]}$$

$$= \cos\theta + i\sin\theta \text{ [We used Eq. (12) and Eq. (13)]}$$

Hence, $e^{i\theta} = \cos\theta + i\sin\theta$ .... Eq. (14).

Eq. (14), known as Euler's identity, relates $e^{i\theta}$ to the sine and cosine functions: an exponential function is related to the trig-ratios through the mediation of the imaginary number i. Amazing!

Chapter 1: Mathematical Prerequisites

Integrals with Complex Numbers

In this section, our goal is to carry out Gaussian integrals (the cornerstone of Path Integrals) involving the imaginary number i. As a step toward that, we will evaluate the integral of e^{ix}. This is to give you confidence that integrals of Complex numbers are very much like those involving real numbers.

In carrying out $\int e^{ix} dx$, we will write e^{ix} in terms of its real and imaginary parts and then integrate them separately.

$\int e^{ix} dx$
$= \int dx [\cos x + i \sin x]$ [Using Euler's identity given by Eq. (14).]
$= \int dx \cos x + i \int dx \sin x$ [i is treated as a constant]
$= \sin x - i \cos x + C$ [We used $\int dx \cos x = \sin x$ and
$\int dx \sin x = -\cos x$. C is a constant]
$= \sin x - i \cos x + C$ .... Eq. (15).

Next, we will derive the above result differently: by substituting $u = ix$ in the integral $\int e^{ix} dx$, and treating i as a constant.

Since $u = ix$, $du = idx \Rightarrow dx = \frac{du}{i}$.

Hence, $\int e^{ix} dx$

$= \int e^u \frac{du}{i}$ [Using $dx = \frac{du}{i}$]
$= \frac{1}{i} \int e^u du$
$= \frac{1}{i} e^u + C$ [We used $\int e^u du = e^u$]
$= \frac{1}{i} e^{ix} + C$ [Substituting $u = ix$]
$= -i e^{ix} + C$ [Using $\frac{1}{i} = -i$]
$= -i(\cos x + i \sin x) + C$ [Using Eq. (14)]
$= -i \cos x + \sin x + C$. This is precisely what we got in Eq. (15).

Feynman's Path Integral explained with basic Calculus

In the above example, you saw that the imaginary number i did not pose any additional mathematical challenge in evaluating an integral. In an integral, you can treat i as a constant having the property that its square is negative 1: $i^2 = -1$.

Gaussian Integrals with Imaginary Numbers

Next, consider the Gaussian integral: $\int_{-\infty}^{+\infty} e^{ix^2} dx$. You might say, wait a minute, this is just like the integral $\int_{-\infty}^{+\infty} e^{-Ax^2+Bx} dx$ we saw in Eq. (6), with $B = 0$ and $A = -i$. (Never mind that A and B in Eq. (6) are real, with A being greater than zero.) Using the result of Eq. (6), viz., $\int_{-\infty}^{+\infty} e^{-Ax^2+Bx} dx = \sqrt{\frac{\pi}{A}} e^{\frac{B^2}{4A}}$

with $A = -i$ and $B = 0$, we obtain $\int_{-\infty}^{+\infty} e^{ix^2} dx = \sqrt{\frac{\pi}{A}} e^{\frac{B^2}{4A}} = \sqrt{\frac{\pi}{-i}} = \sqrt{i\pi}$

(Using $\frac{1}{-i} = i$). And guess what, this is correct!

Pause and think what we just did! We calculated $\int_{-\infty}^{+\infty} e^{ix^2} dx$ by using a formula that we had derived for real A and B. We simply pretended that $\int_{-\infty}^{+\infty} e^{-Ax^2+Bx} dx = \sqrt{\frac{\pi}{A}} e^{\frac{B^2}{4A}}$ is valid even when A and B are complex and we used an imaginary number for A. In fact, our guess is correct; for the kind of integrals we use in this book, Eq. (6) does turn out to be true, even when A and B are complex and a finite value for the integral exists.

The discussion we just had is reminiscent of how $(1 + x)^n = 1 + nx + \frac{n(n-1)}{2!} x^2 + \frac{n(n-1)(n-2)}{3!} x^3 + \cdots$ is valid not only for integer values of n, but for all n.

Chapter 1: Mathematical Prerequisites

A Word on Convergence

Why does the integral $\int_{-\infty}^{+\infty} e^{ix^2} dx$ converge? The question arises since, per Eq. (14), $e^{ix^2} = \cos x^2 + i \sin x^2$. Hence, evaluating $\int_{-\infty}^{+\infty} dx\, e^{ix^2}$ means you integrate the sine and cosine functions from minus infinity to plus infinity, which will give non-convergent results since sines and cosines are oscillatory. But we obtained a finite value for $\int_{-\infty}^{+\infty} dx\, e^{ix^2}$. So, what is going on?

You can think of $\int_{-\infty}^{+\infty} dx\, e^{ix^2}$ as $\int_{-\infty}^{+\infty} dx\, e^{-\epsilon x^2} e^{ix^2}$, where ϵ is infinitesimally small. We justify this step by noting that the two integrands e^{ix^2} and $e^{-\epsilon x^2} e^{ix^2}$ are the same in the limit $\epsilon \to 0$. After evaluating $\int_{-\infty}^{+\infty} dx\, e^{-\epsilon x^2} e^{ix^2}$ (which is guaranteed to converge due to the presence of the damping factor $e^{-\epsilon x^2}$, which is way smaller in magnitude than the other factor for large $|x|$), we will take the limit $\epsilon \to 0$ in the final result.

$\int_{-\infty}^{+\infty} dx\, e^{-\epsilon x^2} e^{ix^2} = \int_{-\infty}^{+\infty} dx\, e^{-(\epsilon - i)x^2}$, which we can evaluate using Eq. (6): $\int_{-\infty}^{+\infty} e^{-Ax^2 + Bx} dx = \sqrt{\frac{\pi}{A}} e^{\frac{B^2}{4A}}$, with $A = (\epsilon - i)$ and $B = 0$. So,

$\sqrt{\frac{\pi}{A}} e^{\frac{B^2}{4A}} = \sqrt{\frac{\pi}{\epsilon - i}} = \sqrt{\frac{\pi}{-i}}$. (Taking the limit $\epsilon \to 0$). Hence, $\int_{-\infty}^{+\infty} dx\, e^{-\epsilon x^2} e^{ix^2} = \sqrt{\frac{\pi}{-i}} = \sqrt{i\pi}$, same as what we got previously.

RECAP: $\int_{-\infty}^{+\infty} e^{-Ax^2 + Bx} dx = \sqrt{\frac{\pi}{A}} e^{\frac{B^2}{4A}}$ is valid even when A and B are complex, provided a finite value for the integral exists.

Feynman's Path Integral explained with basic Calculus

Additional Exercises

1. Prove

$$\int_{-\infty}^{+\infty} dx e^{[-k_1(x-a)^2 - k_2(x-b)^2]} = \sqrt{\frac{\pi}{k_1+k_2}} e^{\frac{-k_1 k_2}{k_1+k_2}(a-b)^2} \quad \ldots\ldots\ldots \text{Eq. (16)}.$$

Hint. First write $-k_1(x-a)^2 - k_2(x-b)^2$ in the form: $Ax^2 + Bx + C$ by expanding each of the squared terms. Then use Eq. (6).

Note. Eq. (16) holds even when k is an imaginary number. (We should be comfortable with the idea of an expression being valid for both real and complex numbers by now.) Plugging in $k_1 = -ip_1$ and $k_2 = -ip_2$ in Eq. (16), we get:

$$\int_{-\infty}^{+\infty} dx e^{i[p_1(x-a)^2 + p_2(x-b)^2]} = \sqrt{\frac{i\pi}{p_1+p_2}} e^{\frac{ip_1 p_2}{p_1+p_2}(a-b)^2} \quad \ldots\ldots\ldots \text{Eq. (17)}.$$

We will need equations 16 and 17 for carrying out Path Integrals.

2. Integrate $\int_{-\infty}^{+\infty} \int_{-\infty}^{+\infty} dx dy e^{i[(x-a)^2 + (x-y)^2 + (y-b)^2]}$

Hint. Solve the above as iterated integrals, i.e., first, carry out the integral w.r.t y, treating x as a constant. Then do the x-integral. For example,

$$\int_{-\infty}^{+\infty} \int_{-\infty}^{+\infty} dx dy e^{i[(x-a)^2 + (x-y)^2 + (y-b)^2]}$$

$$= \int_{-\infty}^{+\infty} dx e^{i(x-a)^2} \int_{-\infty}^{+\infty} dy e^{i[(x-y)^2 + (y-b)^2]}$$

The y-integral in the above can be carried out using Eq. (17). Then, the x-integral can be carried out similarly.

3. <u>Advanced Problem</u> Prove that:

$$\frac{1}{\sigma\sqrt{2\pi}} \int_{-\infty}^{+\infty} dx x^2 e^{-\frac{x^2}{2\sigma^2}} = \sigma^2 \ldots\ldots\ldots\ldots\ldots\ldots\ldots\ldots \text{Eq. (18)}.$$

17

Chapter 1: Mathematical Prerequisites

You will need Eq. (18) in Chapter 6 while deriving Schrodinger's equation. You can either memorize Eq. (18) and move on or go over its derivation given below. The derivation will require you to carry out differentiation inside an integral, which might be a bit intimidating if you haven't seen it before. But it could also be fun, depending on your perspective.

<u>Soln.</u> We will first evaluate $\phi(k) = \int_{-\infty}^{+\infty} dx\, e^{ikx} e^{-\frac{x^2}{2\sigma^2}}$ (with the help of Eq. (6)), because the integral we are trying to evaluate is related to $\phi(k)$ through the following identity (the proof follows shortly)

$$\left.\frac{d^2\phi(k)}{dk^2}\right|_{k=0} = -\int_{-\infty}^{+\infty} dx\, x^2 e^{-\frac{x^2}{2\sigma^2}} \quad \ldots \ldots \ldots \ldots \ldots \ldots \text{Eq. (19).}$$

So, if we can evaluate $\phi(k)$, its second derivative at $k = 0$ will lead us to the sought-after integral of Eq. (18). First, we prove Eq. (19).

Proof of Eq. (19):

Since $\phi(k) = \int_{-\infty}^{+\infty} dx\, e^{ikx} e^{-\frac{x^2}{2\sigma^2}}$,

$\frac{d^2\phi(k)}{dk^2} = \frac{d^2}{dk^2} \int_{-\infty}^{+\infty} dx\, e^{ikx} e^{-\frac{x^2}{2\sigma^2}}$, which can be evaluated by differentiating under the integral sign as follows:

$\frac{d^2\phi(k)}{dk^2} = \int_{-\infty}^{+\infty} dx \left(\frac{\partial^2}{\partial k^2} e^{ikx}\right) e^{-\frac{x^2}{2\sigma^2}}$. [$\frac{\partial}{\partial k}$ is $\frac{d}{dk}$ with x treated as a constant. Since $\frac{d}{dk}$ does not depend on x, we moved $\frac{d}{dk}$ from outside an integral to the inside.]

$= \int_{-\infty}^{+\infty} dx (-x^2 e^{ikx}) e^{-\frac{x^2}{2\sigma^2}}$

[We used $\frac{\partial^2}{\partial k^2} e^{ikx} = (ix)^2 e^{ikx} = -x^2 e^{ikx}$]

$= -\int_{-\infty}^{+\infty} (x^2 e^{ikx}) e^{-\frac{x^2}{2\sigma^2}}$

Feynman's Path Integral explained with basic Calculus

$$\Rightarrow \frac{d^2\phi(k)}{dk^2}\bigg|_{k=0} = -\int_{-\infty}^{+\infty} dx\, x^2 e^{ikx}\bigg|_{k=0} e^{-\frac{x^2}{2\sigma^2}}$$

$$= -\int_{-\infty}^{+\infty} dx\, x^2 e^{-\frac{x^2}{2\sigma^2}} \quad [\text{Since } e^{ikx}\big|_{k=0} = 1]$$

(Eq. (19) Proved)).

Now that we established Eq. (19), we will evaluate $\phi(k) = \int_{-\infty}^{+\infty} dx\, e^{ikx} e^{-\frac{x^2}{2\sigma^2}}$. Setting $A = \frac{1}{2\sigma^2}$, and $B = ik$ in Eq. (6), we get, $\phi(k) = \sigma\sqrt{2\pi}\, e^{-\frac{\sigma^2 k^2}{2}}$, which gives $\frac{d^2\phi(k)}{dk^2} = \sigma^3\sqrt{2\pi}\, e^{-\frac{\sigma^2 k^2}{2}}[\sigma^2 k^2 - 1]$. Evaluating $\frac{d^2\phi(k)}{dk^2}$ at $k = 0$, we get $\frac{d^2\phi(k)}{dk^2}\bigg|_{k=0} = -\sigma^3\sqrt{2\pi}$, using which in Eq. (19), we get, $\int_{-\infty}^{+\infty} dx\, x^2 e^{-\frac{x^2}{2\sigma^2}} = \sigma^3\sqrt{2\pi}$. Next dividing both sides by $\sigma\sqrt{2\pi}$, we obtain Eq. (18). (Proved).

RECAP: Important formulae involving Gaussian integrals:

1. $\int_{-\infty}^{+\infty} e^{-Ax^2+Bx}\, dx = \sqrt{\frac{\pi}{A}}\, e^{\frac{B^2}{4A}}$

2. $\int_{-\infty}^{+\infty} dx\, e^{[-k_1(x-a)^2 - k_2(x-b)^2]} = \sqrt{\frac{\pi}{k_1+k_2}}\, e^{\frac{-k_1 k_2}{k_1+k_2}(a-b)^2}$

3. $\int_{-\infty}^{+\infty} dx\, e^{i[p_1(x-a)^2 + p_2(x-b)^2]} = \sqrt{\frac{i\pi}{p_1+p_2}}\, e^{\frac{ip_1 p_2}{p_1+p_2}(a-b)^2}$

A function f(x) can be Taylor expanded about x = a as follows:

$$f(x) = f(x)\big|_{x=a} + \frac{df}{dx}\bigg|_{x=a}(x-a) + \frac{1}{2!}\frac{d^2 f}{dx^2}\bigg|_{x=a}(x-a)^2 + \frac{1}{3!}\frac{d^3 f}{dx^3}\bigg|_{x=a}(x-a)^3 + \cdots$$

Chapter 2: Variational Calculus

In this chapter, we will talk about Variational Calculus, which, as you will see later, makes Path Integrals easy to evaluate. Variational Calculus also plays a key role in establishing the connection between the classical and quantum mechanics via the Path Integral formalism.

We introduce you to the concept of Variational Calculus through an example. We know that the shortest path between two points is the straight line joining the points. We will show you how you can prove that, by using concepts from Variational Calculus or the Calculus of Variations, as it is also called. As we walk you through the proof, you will grasp the core principle.

The minimum of a single-variable function $f(x)$ occurs at an x where the derivative of the function $\frac{df}{dx}$ is 0 (See Figure 2). Equivalently, a function is *stationary* near where it assumes a minimum (or a maximum), i.e., its *first order change* is 0 (in other words, the function is roughly a constant). This can be seen from $df = \frac{df}{dx}dx$, where df is the first order change in the function. Clearly, $df = 0$, when $\frac{df}{dx} = 0$, i.e., the function f assumes its minimum value.

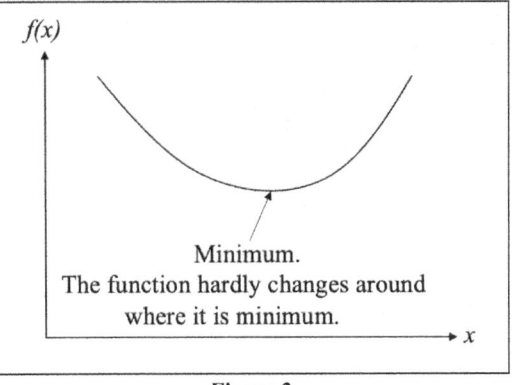

Figure 2: Minimum. The function hardly changes around where it is minimum.

In a similar vein, we will show that the length of a straight line joining two points is "stationary". This means that if we produce a new curve by adding a small variation to the straight line, the length of the new curve will be the same

Feynman's Path Integral explained with basic Calculus

as that of the straight line (up to the first order!) This could not be true if the straight line was not the shortest path between the points, which is "analogous" to df not being 0 when $\frac{df}{dx} \neq 0$, i.e., when $f(x)$ is not at its minimum.

Assume that a straight line given by the equation $\bar{y}(x) = mx$ joins the origin $O(0,0)$ and an arbitrary point $A(a, am)$ (See Figure 3). We change the straight-line slightly by adding a small but arbitrary function $\eta(x)$ to it, to produce a new function $y(x) = \bar{y}(x) + \eta(x)$ such that the endpoints remain the same (O and A). This implies that $\eta(0) = \eta(a) = 0$ as shown in Figure 3.

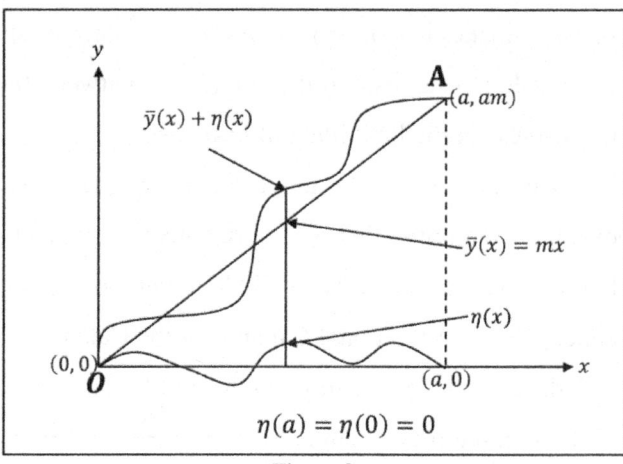

Figure 3

We will show that the length of the curve $y(x) = \bar{y}(x) + \eta(x)$ between O and A in Figure 3 is the same as that of the straight line OA (to the first order).

Using jargon from Calculus of Variations, the length is a "functional", whose value depends on a function over an interval. As we know, a function $f(x)$ is an input-output system (See Figure 4) that takes a number x as the input and spits out another number $f(x)$ as the output. However, the input to a *functional* is a whole function and the output is just a number (See Figure 5). For example, the "length-functional" of a curve $f(x)$ depends not just on one value of $f(x)$ but on the entire $f(x)$ defined over an interval of x. A functional is sometimes denoted by the symbol

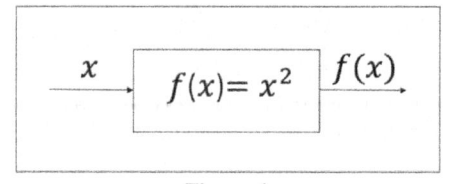

Figure 4

21

Chapter 2: Variational Calculus

$F[f(x)]$, indicating that for a given function $f(x)$ you get only one number ($F[f(x)]$) as the output. [For more on functionals, see the Appendix 1.]

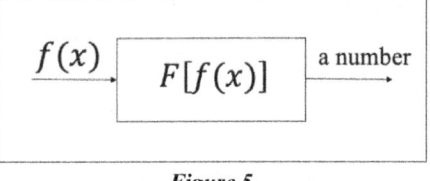

Figure 5

The core concept of Calculus of Variations is to find a function which will minimize a certain functional. However, we turn the problem on its head. We think that the function that minimizes the *length-functional* between two points is a straight line. Let us now formally verify it. To that end, we will calculate the first order change in the length of a straight line and show it to be 0.

<u>Verification</u>

The length (S) of an arbitrary curve can be calculated using the Pythagorean theorem for individual small line segments that approximate the curve. (dx, dy and ds in Figure 6 are the two legs and hypotenuse of a right-angled triangle, respectively). Hence,

$$ds = \sqrt{dx^2 + dy^2} = \sqrt{\left(1 + \left(\frac{dy}{dx}\right)^2\right)} dx,$$

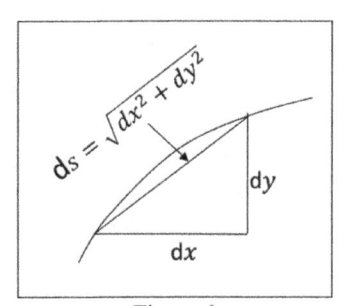

Figure 6

implying

$$S = \int_0^a \sqrt{\left(1 + \left(\frac{dy}{dx}\right)^2\right)} dx \text{ [}x \text{ goes from 0 to } a\text{]} \dots\dots\dots\dots\dots \text{Eq. (1)}.$$

(The following discourse is based on Figure 3, repeated in the following for convenience).

Feynman's Path Integral explained with basic Calculus

For the straight line OA, given by $\bar{y} = mx$, we get $\frac{d\bar{y}}{dx} = m$. Hence the length of the straight line OA, per Eq. (1) is:

$S_{straight-line}$
$= \int_0^a \sqrt{(1 + m^2)} dx$
$= a\sqrt{(1 + m^2)}$ .... Eq. (2).

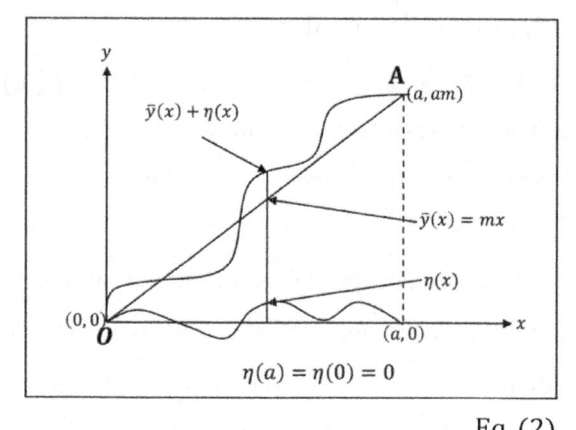

Next, we will find the length for the curve $y(x) \equiv \bar{y}(x) + \eta(x) = mx + \eta(x)$ by using Eq. (1) as follows (Remember $\eta(0) = \eta(a) = 0$).

1. For $y(x) = mx + \eta(x)$, $\frac{dy}{dx} = m + \frac{d\eta}{dx}$ .... Eq. (3).

2. Using Eq. (3) in the formula for the length of a curve given by Eq. (1), we obtain:

$$S_{curve} = \int_0^a \sqrt{1 + \left(m + \frac{d\eta}{dx}\right)^2} dx \quad\ldots\ldots\ldots\ldots\text{Eq. (4).}$$

3. Now, in Eq. (4), $\left(m + \frac{d\eta}{dx}\right)^2 = m^2 + 2m\frac{d\eta}{dx} + \left(\frac{d\eta}{dx}\right)^2$. Neglecting the higher order square term $\left(\frac{d\eta}{dx}\right)^2$, S_{curve} is given as follows (we keep terms up to the ones linear in η and its derivatives, only):

$$S_{curve} \approx \int_0^a \sqrt{\left(1 + m^2 + 2m\frac{d\eta}{dx}\right)} dx$$

$$= \sqrt{(1 + m^2)} \int_0^a \left(1 + \frac{2m\frac{d\eta}{dx}}{1+m^2}\right)^{\frac{1}{2}} dx \quad\ldots\ldots\ldots\ldots\text{Eq. (5).}$$

4. Taylor expanding Eq. (5) using $(1 + u)^{\frac{1}{2}} \approx 1 + \frac{1}{2}u$ (Chapter 1, Eq. (8)), and keeping the lowest order terms, we obtain:

Chapter 2: Variational Calculus

$$S_{curve} \approx \sqrt{(1+m^2)} \int_0^a \left(1 + \frac{1}{2} \cdot \frac{2m\frac{d\eta}{dx}}{1+m^2}\right) dx$$

$$= \sqrt{(1+m^2)}.a + \frac{m}{\sqrt{(1+m^2)}} \int_0^a \frac{d\eta}{dx} dx \ \ldots\ldots\ldots\ldots\ldots\ldots \text{Eq. (6)}.$$

Note that the first term of Eq. (6) is the same as the length of the straight line given by Eq. (2). Hence, from Eq. (6), we obtain:

$$S_{curve} = S_{straight-line} + \frac{m}{\sqrt{(1+m^2)}} \int_0^a \frac{d\eta}{dx} dx \ \ldots\ldots\ldots\ldots\ldots\ldots \text{Eq. (7)}.$$

The integral on the right side of Eq. (7) is: $\int_0^a \frac{d\eta}{dx} dx = \eta(a) - \eta(0) = 0 - 0 = 0$ (η is 0 at the two end points). Hence from Eq. (7) we obtain, $S_{curve} = S_{straight-line}$ to the first order of approximation. In other words, the length of a curve between two given points remains "stationary" (constant/without change) when the curve varies slightly about a straight line joining the two points. This in turn implies that the straight line minimizes the length functional or, in simpler terms, a straight line is shortest path between two given points.

Stationarity Principle

We clearly state the stationarity principle for integrals, which we introduced in our last problem. Suppose you have a definite integral S of a function $\bar{y}(x)$. (S in general, has nothing to do with the arc-length of $\bar{y}(x)$). As you add a small variation to the function $\bar{y}(x)$ keeping the endpoints fixed, you find that the integral S does not change (up to the first order variation). Then, $\bar{y}(x)$ minimizes the integral per the stationarity principle. In other words, the integral S assumes a minimum value for $\bar{y}(x)$, of all functions for which the integral can be calculated.

Conversely, suppose $\bar{y}(x)$ minimizes an integral S. We add a small but arbitrary variation $\eta(x)$ to $\bar{y}(x)$, such that the new curve $y(x) = \bar{y}(x) + \eta(x)$

Feynman's Path Integral explained with basic Calculus

has the same endpoints as the old one. Then, $\bar{y}(x)$ and $y(x)$ will produce the same value for S up to the first order in $\eta(x)$, per the stationarity principle. Let us call $\bar{y}(x)$ as the "least path" (per the language of Variational Calculus, "path" means nothing but a function. The function that minimizes the integral S is the "least path".) The word "least" is used to denote the action being minimized for a certain function. "Least" has nothing to do with an actual "length" or "distance". It is true that the least path for the "arc-length integral" given by Eq. (1) is the shortest path ("least" meaning "shortest"), but the "least paths" for other integrals will not have anything to do with length.

The least path and a neighboring path with same endpoints producing the same S up to the first order of variation, is reminiscent of ordinary differential Calculus where a function $f(x)$ is *stationary* near where it assumes a minimum (See Figure 2 reproduced here for convenience.). At the minimum, you have $\frac{df}{dx} = 0$, which makes *the first order change* in the function: $df = \frac{df}{dx} dx = 0$.

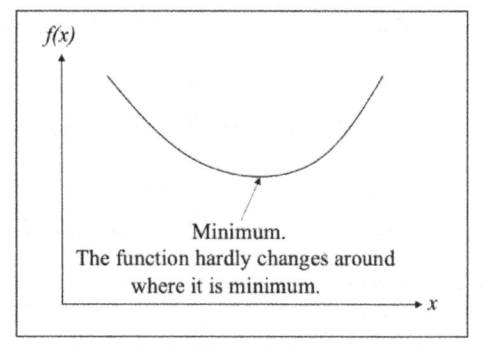

Minimum.
The function hardly changes around where it is minimum.

RECAP: The Stationarity Principle and the Least Path

Suppose $\bar{y}(x)$ minimizes an integral S. We add a small but arbitrary variation $\eta(x)$ to $\bar{y}(x)$, such that the new curve $y(x) = \bar{y}(x) + \eta(x)$ has the same endpoints as the old one. Then, $\bar{y}(x)$ and $y(x)$ will generate the same value for S up to the first order in $\eta(x)$, per the stationarity principle. We refer to $\bar{y}(x)$ as "the least path" (Path means nothing but a function. We call the function that minimizes the integral S as the "least path".)

Chapter 2: Variational Calculus

Deriving the Equation for the Shortest Path

In this section, once again we take up the problem of shortest arc length. We denote $\bar{y}(x)$ to be the "least path", i.e., the function that minimizes the arc length between two points. But unlike the previous section, we will not assume beforehand that $\bar{y}(x)$ is a straight line (Our stance this time is we do not know what the least path, or the shortest curve looks like.) In the following, we will obtain a differential equation for $\bar{y}(x)$ by using the stationarity principle introduced in the last section and show by solving the equation that you get a straight line. Thus, we will establish that a straight line has the minimum distance of all the curves joining two points.

The arc-length of a function $y(x)$ in Figure 7 between the points (x_a, y_a) and (x_b, y_b) is given by:

$$S = \int_{x_a}^{x_b} \sqrt{1 + \left(\frac{dy}{dx}\right)^2} \, dx \quad \ldots\ldots \text{ Eq. (8)}.$$

Let S assume the minimum value S_0 when evaluated for the least path $y(x) = \bar{y}(x)$. Hence:

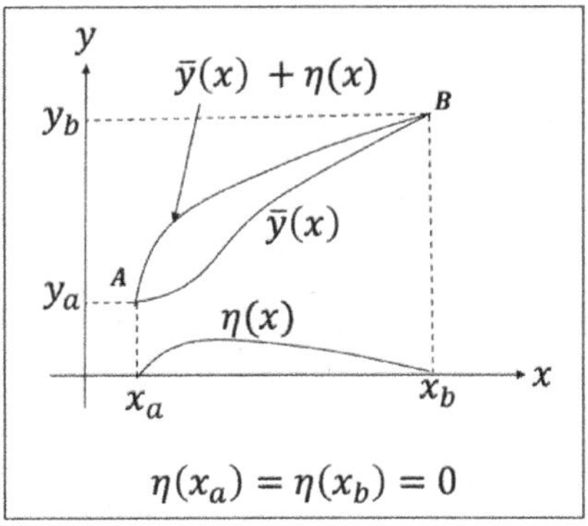

$$S_0 = \int_{x_a}^{x_b} \sqrt{1 + \left(\frac{d\bar{y}}{dx}\right)^2} \, dx \quad \ldots\ldots\ldots\ldots\ldots\ldots\ldots\ldots\ldots\ldots \text{ Eq. (9)}.$$

Figure 7

$\eta(x_a) = \eta(x_b) = 0$

Adding a small but arbitrary variation $\eta(x)$ to $\bar{y}(x)$, we get a new curve $y(x) = \bar{y}(x) + \eta(x)$. Using Eq. (8), the length of the new curve $y(x)$, denoted as S', is given by:

$$S' = \int_{x_a}^{x_b} \sqrt{1 + \left(\frac{d}{dx}(\bar{y}(x) + \eta(x))\right)^2} \, dx \quad \ldots\ldots\ldots\ldots\ldots\ldots \text{Eq. (10)}.$$

Feynman's Path Integral explained with basic Calculus

Since, we consider variations about the least path, the stationarity principle dictates that S' and S_0, as given by Eqs. 10 and 9 respectively, are the same (up to the first order variation in η). We will deploy the stationarity principle shortly. As a first step, we expand S', as given by Eq. (10), about the least path $\bar{y}(x)$ as follows.

$$S' = \int_{x_a}^{x_b} \sqrt{1 + \left(\frac{d\bar{y}}{dx} + \frac{d\eta}{dx}\right)^2}\, dx$$

$$= \int_{x_a}^{x_b} \sqrt{1 + \left(\frac{d\bar{y}}{dx}\right)^2 + 2\left(\frac{d\bar{y}}{dx}\right)\left(\frac{d\eta}{dx}\right) + \left(\frac{d\eta}{dx}\right)^2}\, dx$$

[We used $(a+b)^2 = a^2 + 2ab + b^2$]

$$\approx \int_{x_a}^{x_b} \sqrt{1 + \left(\frac{d\bar{y}}{dx}\right)^2 + 2\left(\frac{d\bar{y}}{dx}\right)\left(\frac{d\eta}{dx}\right)}\, dx$$

[Ignoring the higher order term $\left(\frac{d\eta}{dx}\right)^2$. We only keep terms linear in $\eta(x)$ and its derivatives.]

$$= \int_{x_a}^{x_b} \sqrt{\left(1 + \left(\frac{d\bar{y}}{dx}\right)^2\right)\left[1 + \frac{2\left(\frac{d\bar{y}}{dx}\right)\left(\frac{d\eta}{dx}\right)}{1+\left(\frac{d\bar{y}}{dx}\right)^2}\right]}\, dx$$

$$= \int_{x_a}^{x_b} \sqrt{\left(1 + \left(\frac{d\bar{y}}{dx}\right)^2\right)}\left[1 + \frac{2\left(\frac{d\bar{y}}{dx}\right)\left(\frac{d\eta}{dx}\right)}{1+\left(\frac{d\bar{y}}{dx}\right)^2}\right]^{\frac{1}{2}} dx \quad\ldots\ldots\ldots\ldots\ldots\ldots \text{Eq. (11)}.$$

Note that in the above expression, $\frac{2\left(\frac{d\bar{y}}{dx}\right)\left(\frac{d\eta}{dx}\right)}{1+\left(\frac{d\bar{y}}{dx}\right)^2}$ (which we will call z) is much smaller than 1 because of $\frac{d\eta}{dx}$ in the numerator (η is a small variation; η and its derivatives are considered small). Hence, using the Taylor expansion (Chapter 1, Eq. (8)) $(1+z)^{\frac{1}{2}} \approx 1 + \frac{1}{2}z$ in Eq. (11)), we get:

$$S' \approx \int_{x_a}^{x_b} \sqrt{\left(1 + \left(\frac{d\bar{y}}{dx}\right)^2\right)}\left[1 + \frac{\left(\frac{d\bar{y}}{dx}\right)\left(\frac{d\eta}{dx}\right)}{1+\left(\frac{d\bar{y}}{dx}\right)^2}\right] dx$$

Chapter 2: Variational Calculus

$$= \int_{x_a}^{x_b} \sqrt{1 + \left(\frac{d\bar{y}}{dx}\right)^2}\, dx + \int_{x_a}^{x_b} \frac{\left(\frac{d\bar{y}}{dx}\right)\left(\frac{d\eta}{dx}\right)}{\sqrt{1 + \left(\frac{d\bar{y}}{dx}\right)^2}}\, dx \quad \ldots\ldots\ldots\ldots\ldots\text{Eq. (12).}$$

The first term on the right of Eq. (12) is S_0 (given by Eq. (9)), which we will set equal to S' on the left of Eq. (12), by invoking the previously mentioned stationarity principle. Setting $S' = S_0$ in Eq. (12), we get

$$\int_{x_a}^{x_b} \frac{\left(\frac{d\bar{y}}{dx}\right)}{\sqrt{1 + \left(\frac{d\bar{y}}{dx}\right)^2}} \left(\frac{d\eta}{dx}\right)\, dx = 0 \quad \ldots\ldots\ldots\ldots\ldots\ldots\ldots\ldots\text{Eq. (13).}$$

Setting $\dfrac{\left(\frac{d\bar{y}}{dx}\right)}{\sqrt{1 + \left(\frac{d\bar{y}}{dx}\right)^2}} = u(x)$, $\left(\dfrac{d\eta}{dx}\right) dx = dv$ in Eq. (13) and doing integration by parts ($\int u\, dv = uv - \int v\, du$), we obtain the following from Eq. (13):

$$\left. \frac{\left(\frac{d\bar{y}}{dx}\right)}{\sqrt{1 + \left(\frac{d\bar{y}}{dx}\right)^2}} \eta(x) \right|_{x=x_a}^{x=x_b} - \int_{x_a}^{x_b} dx\, \eta(x) \frac{d}{dx}\left(\frac{\left(\frac{d\bar{y}}{dx}\right)}{\sqrt{1 + \left(\frac{d\bar{y}}{dx}\right)^2}} \right) = 0 \ldots \text{Eq. (14).}$$

But, since $\eta(x)$ is 0 at $x = x_a$ and $x = x_b$, the first term in Eq. (14) vanishes. Hence, Eq. (14) produces:

$$\int_{x_a}^{x_b} \eta(x) \frac{d}{dx}\left(\frac{\left(\frac{d\bar{y}}{dx}\right)}{\sqrt{1 + \left(\frac{d\bar{y}}{dx}\right)^2}} \right) = 0.$$

Now, the above equation is true for any arbitrary $\eta(x)$, which means:

$$\frac{d}{dx}\left(\frac{\left(\frac{d\bar{y}}{dx}\right)}{\sqrt{1 + \left(\frac{d\bar{y}}{dx}\right)^2}} \right) = 0 \quad \ldots\ldots\ldots\ldots\ldots\ldots\ldots\ldots\ldots\text{Eq. (15).}$$

We will show that you get a straight line by solving Eq. (15) for $\bar{y}(x)$. Before we actually do that, you can see that is the case by setting $\dfrac{d\bar{y}}{dx} = a\ constant$

28

Feynman's Path Integral explained with basic Calculus

on the left side of Eq. (15). If $\frac{d\bar{y}}{dx}$ is a constant, the left side of Eq. (15) vanishes, in agreement with what you have on the right. But $\frac{d\bar{y}}{dx}$ can only be a constant if $\bar{y}(x)$ represents a straight line.

Now, the formal derivation. Integrating both sides of Eq. (15) w.r.t x, we obtain:

$$\frac{\left(\frac{d\bar{y}}{dx}\right)}{\sqrt{1+\left(\frac{d\bar{y}}{dx}\right)^2}} = C_1,$$

where C_1 is a constant. Defining $\frac{d\bar{y}}{dx} = m$, we have

$\frac{m}{\sqrt{1+m^2}} = C_1$, squaring both sides of which, we get $\frac{m^2}{(1+m^2)} = C_1^2$. Solving for m, we will get $m = \frac{C_1}{\sqrt{1-C_1^2}} \equiv K$, a constant. But m is $\frac{d\bar{y}}{dx}$, implying: $\frac{d\bar{y}}{dx} = K$.

Solving, we get $\bar{y}(x) = Kx + Q$ (both K and Q are constants), which is clearly the equation of a straight line. You can determine K and Q, by using the given values of $\bar{y}(x)$ at $x = x_a$ and $x = x_b$. (See the Exercise below.)

Thus, we derived the differential equation of the curve of shortest length between two points and discovered by solving the equation that the curve is a straight line.

Exercise 1. Starting with $\bar{y}(x) = Kx + Q$ and using $\bar{y}(x = x_a) = y_a$ and $\bar{y}(x = x_b) = y_b$, find K and Q.

RECAP: How to obtain a differential equation for the least path?

- Expand the integral in question about the least path.
- Only retain terms linear in η (the variation about the least path).
- Use the stationarity principle to obtain a differential equation, the solution to which gives the least path.

Chapter 2: Variational Calculus

Further Examples of Variational Calculus

In the following, we will show you three interesting examples from Variational Calculus. In each case, we will give you an integral and you will find the least path, i.e., the function that minimizes the integral. First you will obtain a differential equation for the least path using the stationarity principle, then solve the equation to obtain the least path. Later, we will use these examples to explain a deep physical principle.

Example 1. Consider the following integral:

$$S = \int_{t=t_a}^{t=t_b} \frac{m}{2}\left(\frac{dy(t)}{dt}\right)^2 dt \quad \dots\dots\dots\dots\dots\dots\dots\dots \text{Eq. (16).}$$

m in Eq. (16) is a constant. (Do not worry about why we have a factor of $\frac{1}{2}$ in the integral for now. Treat it as a numerical constant.)

Find $y(t)$ that minimizes S in Eq. (16). The values of $y(t)$ at the endpoints $t = t_a$ and $t = t_b$ are fixed: $y(t = t_a) = y_a$ and $y(t = t_b) = y_b$.

<u>Soln.</u> Let, of all the curves joining the points A and B shown in Figure 8, $y(t) = \bar{y}(t)$ represent the least path, for which the integral S assumes its minimum value S_0. Hence:

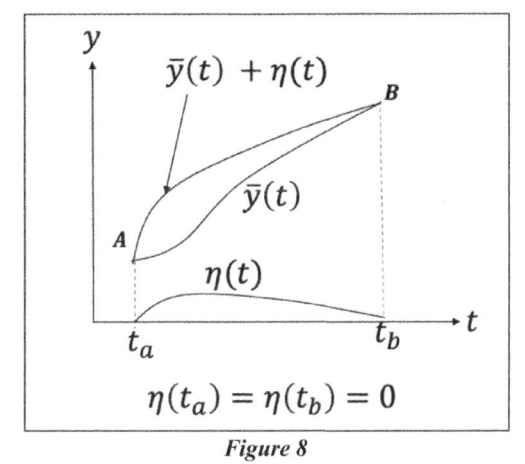

Figure 8

$\bar{y}(t) + \eta(t)$
$\bar{y}(t)$
$\eta(t)$
$\eta(t_a) = \eta(t_b) = 0$

$$S_0 = \int_{t=t_a}^{t=t_b} \frac{m}{2}\left(\frac{d\bar{y}(t)}{dt}\right)^2 dt \quad \dots\dots\dots\dots\dots\dots\dots\dots \text{Eq. (17).}$$

Feynman's Path Integral explained with basic Calculus

Our goal is to find a differential equation for $\bar{y}(t)$, solving which you get the least path for the integral given by Eq. (16).

As in the previous problem, we add a small but arbitrary variation $\eta(t)$ to $\bar{y}(t)$ to get a new curve $y(t) = \bar{y}(t) + \eta(t)$. The integral corresponding to the new curve $y(t)$, denoted as S' is given by:

$$S' = \int_{t=t_a}^{t=t_b} \frac{m}{2} \left(\frac{d}{dt}(\bar{y}(t) + \eta(t))\right)^2 dt \quad \text{........................ Eq. (18).}$$

Also, similar to the previous problem, the idea is to expand S' in Eq. (18), keep terms up to those linear in η and its derivatives, and set the resulting expression equal to S_0 (given by Eq. (17)). (Since, we consider variations about the least path, the stationarity principle dictates that S' and S_0 are the same (up to the first order variation in η).)

Expanding S', given by Eq. (18), we get:

$$S' = \frac{m}{2} \int_{t=t_a}^{t=t_b} \left(\frac{d}{dt}(\bar{y}(t) + \eta(t))\right)^2 dt$$

$$= \frac{m}{2} \int_{t=t_a}^{t=t_b} \left(\frac{d\bar{y}}{dt} + \frac{d\eta}{dt}\right)^2 dt$$

$$= \frac{m}{2} \int_{t=t_a}^{t=t_b} \left(\frac{d\bar{y}}{dt}\right)^2 dt + m \int_{t=t_a}^{t=t_b} \left[\frac{d\bar{y}}{dt}\frac{d\eta}{dt}\right] + \frac{m}{2} \int_{t=t_a}^{t=t_b} \left(\frac{d\eta}{dt}\right)^2 dt$$

[We used $(a+b)^2 = a^2 + 2ab + b^2$]

$$\approx \frac{m}{2} \int_{t=t_a}^{t=t_b} \left(\frac{d\bar{y}}{dt}\right)^2 dt + m \int_{t=t_a}^{t=t_b} \left[\frac{d\bar{y}}{dt}\frac{d\eta}{dt}\right] dt \quad \text{[We ignored the higher}$$

order term $\left(\frac{d\eta}{dt}\right)^2$, since η and its

derivatives are small]

.... Eq. (19).

We carry out integration by parts on the term $\int_{t=t_a}^{t=t_b} \frac{d\bar{y}}{dt}\frac{d\eta}{dt} dt$. With $u = \frac{d\bar{y}}{dt}$, and $dv = \frac{d\eta}{dt} dt$, and using $\int u\,dv = uv - \int v\,du$, we obtain:

$$\int_{t=t_a}^{t=t_b} \frac{d\bar{y}}{dt}\frac{d\eta}{dt} dt = \frac{d\bar{y}}{dt}\eta(t)\Big|_{t=t_a}^{t=t_b} - \int_{t=t_a}^{t=t_b} \eta(t)\frac{d}{dt}\left(\frac{d\bar{y}}{dt}\right) dt \quad \text{...... Eq. (20).}$$

Chapter 2: Variational Calculus

Now, since $\eta(t_b) = \eta(t_a) = 0$, the first term on the right of Eq. (20) vanishes. Hence, from Eq. (20), we get:

$\int_{t=t_a}^{t=t_b} \frac{d\bar{y}}{dt} \frac{d\eta}{dt} dt = -\int_{t=t_a}^{t=t_b} \eta(t) \frac{d^2\bar{y}}{dt^2} dt$, using which in Eq. (19), we get:

$$S' = \frac{m}{2} \int_{t=t_a}^{t=t_b} \left(\frac{d\bar{y}}{dt}\right)^2 dt - m \int_{t=t_a}^{t=t_b} \eta(t) \frac{d^2\bar{y}}{dt^2} dt$$

But the first term in the above is simply S_0 (given by Eq. (17)). Hence, we obtain:

$$S' = S_0 - m \int_{t=t_a}^{t=t_b} \eta(t) \frac{d^2\bar{y}}{dt^2} dt \quad \ldots \ldots \text{Eq. (21).}$$

But $S' = S_0$, per the stationarity principle; hence from Eq. (21), $\int_{t=t_a}^{t=t_b} \eta(t) \frac{d^2\bar{y}}{dt^2} dt = 0$, which implies that $\frac{d^2\bar{y}}{dt^2} = 0$, $\eta(t)$ being arbitrary. Hence, the differential equation that the least path $\bar{y}(t)$ satisfies is:

$$\frac{d^2\bar{y}}{dt^2} = 0 \quad \ldots \ldots \text{Eq. (22).}$$

Integrating $\frac{d^2\bar{y}}{dt^2} = 0$ twice, you obtain $\bar{y}(t) = Kt + Q$, which is the equation of a straight line (K and Q are constants.). So, a straight line (Figure 9) minimizes the integral in Eq. (16), viz., $S = \int_{t=t_a}^{t=t_b} \frac{m}{2} \left(\frac{dy(t)}{dt}\right)^2 dt$, where $y(t)$ assumes fixed values at the end-points $t = t_a$ and $t = t_b$.

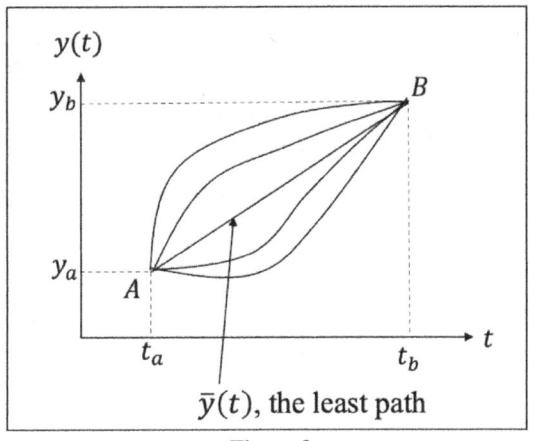

Figure 9

We began this chapter by proving that a straight line minimizes the arc length of a curve (expressed as an integral) between two points. In this example, we found that a straight line minimizes yet another integral.

Feynman's Path Integral explained with basic Calculus

Exercise 2. As we discussed, the solution to Eq. (22) is given by $\bar{y}(t) = Kt + Q$. Find K and Q, given that $\bar{y}(t = t_a) = y_a$ and $\bar{y}(t = t_b) = y_b$. Then evaluate the minimum value for the integral $S = \int_{t=t_a}^{t=t_b} \left[\frac{1}{2} m \left(\frac{d\bar{y}}{dt} \right)^2 \right] dt$, i.e., calculate S for the least path per Eq. (17): $S_0 = \int_{t=t_a}^{t=t_b} \left[\frac{1}{2} m \left(\frac{d\bar{y}}{dt} \right)^2 \right] dt$.

Ans. $S_0 = \frac{m}{2} \frac{(y_b - y_a)^2}{t_b - t_a}$ .... Eq. (23).

Example 2. In this example, we consider the following integral:

$$S = \int_{t=t_a}^{t=t_b} \left[\frac{m}{2} \left(\frac{dy(t)}{dt} \right)^2 + mgy(t) \right] dt \quad \text{............................ Eq. (24).}$$

As you can see, in Eq. (24), we have a new term $mgy(t)$ as compared to Eq. (16) (in Example 1). m and g is are constants.

We would like to find a $y(t)$ that minimizes S, given by Eq. (24). We denote the least path as $\bar{y}(t)$, i.e., S assumes its minimum value S_0 when $y(t) = \bar{y}(t)$. Hence, writing $y(t) = \bar{y}(t)$ in Eq. (24), we get:

$$S_0 = \int_{t=t_a}^{t=t_b} \left[\frac{m}{2} \left(\frac{d\bar{y}}{dt} \right)^2 + mg\bar{y} \right] dt \quad \text{...................................... Eq. (25).}$$

Let S' correspond to $y(t) = \bar{y}(t) + \eta(t)$, where $\eta(t)$ is a small variation about $\bar{y}(t)$. Hence:

$$S' = \int_{t=t_a}^{t=t_b} \left[\frac{m}{2} \left(\frac{d}{dt} (\bar{y} + \eta) \right)^2 + mg(\bar{y} + \eta) \right] dt \quad \text{................ Eq. (26).}$$

As in the previous examples, the idea is to expand S', given by the above expression, only upto terms linear in η and its derivatives, and then set S' equal to S_0 (given by Eq. (25)), by invoking the stationarity principle; finally, obtain a differential equation for the least path $\bar{y}$.

Expanding Eq. (26) about $\bar{y}$,

$$S' = \int_{t=t_a}^{t=t_b} \left[\frac{m}{2} \left(\frac{d}{dt} (\bar{y} + \eta) \right)^2 + mg(\bar{y} + \eta) \right] dt$$

Chapter 2: Variational Calculus

$$= \int_{t=t_a}^{t=t_b} \left[\frac{m}{2}\left(\frac{d\bar{y}}{dt} + \frac{d\eta}{dt}\right)^2 + mg(\bar{y} + \eta) \right] dt$$

$$= \int_{t=t_a}^{t=t_b} \left[\frac{m}{2}\left(\left(\frac{d\bar{y}}{dt}\right)^2 + 2\frac{d\bar{y}}{dt}\frac{d\eta}{dt} + \left(\frac{d\eta}{dt}\right)^2\right) + mg(\bar{y} + \eta) \right] dt$$

[Using $(a+b)^2 = a^2 + 2ab + b^2$]

$$\approx \int_{t=t_a}^{t=t_b} \left[\frac{m}{2}\left(\left(\frac{d\bar{y}}{dt}\right)^2 + 2\frac{d\bar{y}}{dt}\frac{d\eta}{dt}\right) + mg(\bar{y} + \eta) \right] dt \quad \text{[Neglecting the}$$

higher order term $\left(\frac{d\eta}{dt}\right)^2$]

$$= \int_{t=t_a}^{t=t_b} \left[\left(\frac{m}{2}\left(\frac{d\bar{y}}{dt}\right)^2 + mg\bar{y}\right) + m\frac{d\bar{y}}{dt}\frac{d\eta}{dt} + mg\eta \right] dt$$

[Regrouping the terms]

Using $\int_{t=t_a}^{t=t_b} \left[\left(\frac{m}{2}\left(\frac{d\bar{y}}{dt}\right)^2 + mg\bar{y}\right) \right] dt = S_0$ (by Eq. (25)), in the above, we obtain:

$$S' = S_0 + \int_{t=t_a}^{t=t_b} \left[m\frac{d\bar{y}}{dt}\frac{d\eta}{dt} + mg\eta \right] dt \dots\dots\dots\dots\dots\dots \text{ Eq. (27)}.$$

Integrating $\int_{t=t_a}^{t=t_b} \frac{d\bar{y}}{dt}\frac{d\eta}{dt} dt$ in Eq. (27) by parts, we obtain:

$$\int_{t=t_a}^{t=t_b} \frac{d\bar{y}}{dt}\frac{d\eta}{dt} dt = \frac{d\bar{y}}{dt}\eta(t)\Big|_{t=t_a}^{t=t_b} - \int_{t=t_a}^{t=t_b} \eta(t)\frac{d}{dt}\left(\frac{d\bar{y}}{dt}\right) dt$$

Now, since $\eta(t_b) = \eta(t_a) = 0$, the first term on the right of the above equation vanishes, and we are left with:

$\int_{t=t_a}^{t=t_b} \frac{d\bar{y}}{dt}\frac{d\eta}{dt} dt = -\int_{t=t_a}^{t=t_b} \eta(t)\frac{d^2\bar{y}}{dt^2} dt$, using which in Eq. (27), we obtain:

$$S' = S_0 - m\int_{t=t_a}^{t=t_b} \eta(t)\frac{d^2\bar{y}}{dt^2} dt + \int_{t=t_a}^{t=t_b} mg\eta \, dt$$

$$= S_0 - \int_{t=t_a}^{t=t_b} dt\eta(t)\left[m\frac{d^2\bar{y}}{dt^2} - mg\right]$$

But $S' = S_0$ per the requirement of stationarity for variations about the least path (stationary principle). Hence, from the above equation, we get

Feynman's Path Integral explained with basic Calculus

$\int_{t=t_a}^{t=t_b} dt\eta(t) \left[m\frac{d^2\bar{y}}{dt^2} - mg \right] = 0$, which, being true for any arbitrary $\eta(t)$, gives:

$$m\frac{d^2\bar{y}}{dt^2} - mg = 0$$

Or, $m\frac{d^2\bar{y}}{dt^2} = mg$ .... Eq. (28).

The above equation is equivalent to $\frac{d^2\bar{y}}{dt^2} = g$.

Integrating the equation w.r.t t once, we get $\frac{d\bar{y}}{dt} = gt + C_1$. Integrating once more, we get $\bar{y} = \frac{1}{2}gt^2 + C_1 t + C_2$, which is the equation of a parabola. So, the least path is not a straight line this time; instead, it is a quadratic expression in t, i.e., a parabola (Figure 10)!

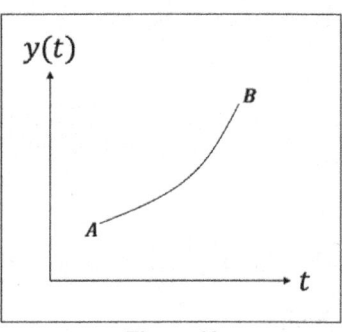

Figure 10

Exercise 3. Find C_1 and C_2 in the equation $\bar{y}(t) = \frac{1}{2}gt^2 + C_1 t + C_2$, given that $\bar{y}(t = t_a) = y_a$ and $\bar{y}(t = t_b) = y_b$. Then evaluate S_0, the minimum value for the integral S given by Eq. (24), i.e., calculate S for the least path $\bar{y}(t)$ per Eq. (25): $S_0 = \int_{t=t_a}^{t=t_b} \left[\frac{1}{2}m \left(\frac{d\bar{y}}{dt}\right)^2 + mg\bar{y} \right] dt$.

Ans. $S_0 = \frac{mg^2}{3}(t_b^3 - t_a^3) + mg(t_b^2 - t_a^2)C_1$

$+ \left(\frac{1}{2}mC_1^2 + mgC_2\right)(t_b - t_a)$ .... Eq. (29),

where, $C_1 = \frac{y_b - y_a}{t_b - t_a} - \frac{1}{2}g(t_a + t_b)$, and $C_2 = \frac{y_a t_b - y_b t_a}{t_b - t_a} + \frac{1}{2}gt_b t_a$.

Chapter 2: Variational Calculus

Physical Significance of S (in the last two examples)

Imagine yourself dropping a stone of mass m from the edge of a cliff. Let $y(t)$ denote the stone's distance from the point it was released at as a function of time t (Figure 11). Since y is the distance and t, the time, $\frac{dy}{dt}$ is the stone's velocity. Hence, $\frac{m}{2}\left(\frac{dy}{dt}\right)^2$ represents the stone's kinetic energy (Kinetic energy $= \frac{1}{2}m(velocity)^2$). The gravitational potential energy (PE) of the stone at the distance y is $PE = -mgy$ (Choosing the downward direction as the positive direction for y). [If you want to brush up your knowledge on potential energy, see Appendix 2.]

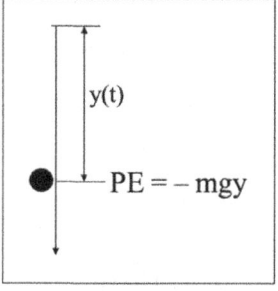

Figure 11

Note that the first term in the integrand of Eq. (24) of the last example (Example 2) is the stone's kinetic energy (KE) and the second term is the negative of the stone's potential energy (PE).

So, per the physical picture we have constructed, Eq. (24) is essentially:

$$S = \int_{t=t_a}^{t=t_b}[KE - PE]dt \quad \text{................................. Eq. (30).}$$

We saw in Eq. (28) that $y(t)$ that minimizes the above S satisfies the differential equation:

$$m\frac{d^2y}{dt^2} = mg \quad \text{.. Eq. (31).}$$

Now, Eq. (31) is precisely Newton's second law of motion for the constant gravitational force ($F = mg$). [Since $\frac{d^2y}{dt^2}$ represents the acceleration a for the distance variable y, the equation $m\frac{d^2y}{dt^2} = mg$ is nothing but Newton's law $F = ma$.]

The above is not a coincidence: there exists a connection between S given by Eq. (30) and Newton's law. S, the time integral of the kinetic energy minus

Feynman's Path Integral explained with basic Calculus

the potential energy, plays a very important role in physical theories including Feynman's Path Integrals, and is called by the name: "action".

To further illustrate the point, we turn our attention to the other example we discussed (Example 1). S in Eq. (16), is the "action" (given by Eq. (30)) with the PE (Potential Energy) term set to 0 (you only have the kinetic energy term $\frac{m}{2}\left(\frac{dy(t)}{dt}\right)^2$ in the integrand). Per Eq. (22), $y(t)$ that minimizes $S = \int_{t=t_a}^{t=t_b} \frac{m}{2}\left(\frac{dy(t)}{dt}\right)^2 dt$, obeys the differential equation: $\frac{d^2y}{dt^2} = 0$. Now, Newton's law for a free particle (not acted upon by an external force) moving along the y-direction looks just the same (The acceleration (represented by $\frac{d^2y}{dt^2}$) being 0 implies the particle is free). It isn't a coincidence that $y(t)$ minimizing an action S with zero PE (i.e. zero external force) also satisfies Newton's law for a free particle, which, once again, shows the connection between the action and Newton's law.

The above discussions are meant to serve as a background for a deep principle of physics called the Least Action Principle.

Least Action Principle

It is a law of nature that a particle, moving under a force field, follows a path that minimizes the integral $S = \int_{t=t_a}^{t=t_b} [KE - PE] dt$, KE and PE being the kinetic and potential energies of the particle respectively. (S is referred as the "action".) Hence, in our parlance of Variational Calculus, a particle's trajectory is the "least path" for the action, (i.e., the path or the function that minimizes the action). And the differential equation that the least path satisfies must be Newton's law of motion, because the least path is the particle's trajectory and the particle's trajectory must obey Newton's law. The particle's trajectory is also called the "classical path". This fact of the classical path

Chapter 2: Variational Calculus

minimizing the corresponding action, and hence being the least path for that action, is called the Least Action Principle.

In the last section, you saw examples of the Least Action Principle (although we didn't use the phrase). In Example 2, we showed that the least path for the action S of a particle (a stone) falling under gravity, with $KE = \frac{1}{2}m\left(\frac{dy}{dt}\right)^2$, and the gravitational $PE = -mgy$, satisfies $m\frac{d^2y}{dt^2} = mg$, which is also Newton's law for an object of mass m falling under the gravitational force mg (i.e., with a constant acceleration g). In Eq. (29), you calculated the action for the stone's trajectory, i.e., its classical path or least path for the corresponding action.

Also, in Example 1 of the same section, we saw that for the action $S = \int_{t=t_a}^{t=t_b}[KE - PE]dt$, with $KE = \frac{1}{2}m\left(\frac{dy}{dt}\right)^2$ and no potential energy, i.e., $PE = 0$, the least path satisfies $m\frac{d^2y}{dt^2} = 0$, which is Newton's law of motion for a "free particle". The action given by Eq. (16) is called the free-particle action and will prove to be immensely useful in understanding Path Integrals. In Eq. (23), you calculated the free-particle-action for the classical path or the least path (the trajectory of the free particle.)

Example 3. You have probably realized that we purposely chose the integrals in Examples 1 and 2 (Eq. (16) and Eq. (24)) so that they were "actions" (with relevance to physical problems). Hence, minimizing those integrals was the same as applying the Least Action Principle. We did not tell you that first, so that you could focus on the math. But that's what we were

Feynman's Path Integral explained with basic Calculus

actually doing. From now on, we will refer to our integrals directly as actions, since only the integrals that are actions are of interest in the book.

Let us now see the third and the final example of the Least Action Principle. Consider a mass m, attached to a spring, oscillating back and forth (Figure 12). The kinetic energy of the system is $KE = \frac{1}{2}m\left(\frac{dx}{dt}\right)^2$. The potential energy of the system (when the spring is stretched by an amount x) is: $PE = \frac{1}{2}kx^2$, where k is the spring constant. Using the above KE and PE in the action given by Eq. (30), we get:

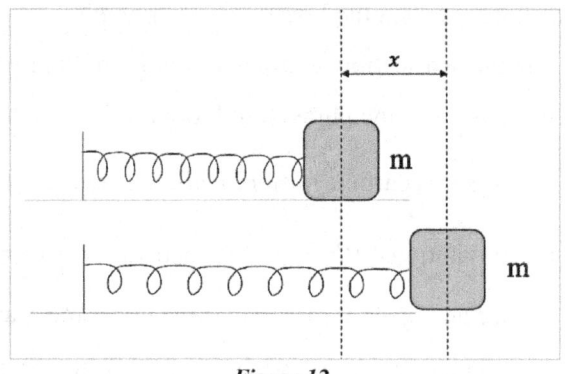

Figure 12

$$S = \int_{t=t_a}^{t=t_b}[KE - PE]dt$$

$$= \int_{t=t_a}^{t=t_b}\left[\frac{1}{2}m\left(\frac{dx}{dt}\right)^2 - \frac{1}{2}kx^2\right]dt \quad\dots\dots\dots\text{ Eq. (32).}$$

Let the action S in Eq. (32) assume its minimum value S_0 when $x(t) = \bar{x}(t)$, i.e., $\bar{x}(t)$ is the least path. Hence,

$$S_0 = \int_{t=t_a}^{t=t_b}\left[\frac{m}{2}\left(\frac{d\bar{x}}{dt}\right)^2 - \frac{1}{2}k\bar{x}^2\right]dt \quad\dots\dots\dots\text{ Eq. (33).}$$

Let S' be the action S corresponding to $x(t) = \bar{x}(t) + \eta(t)$, where $\eta(t)$ is a small variation about $\bar{x}(t)$. Hence,

$$S' = \int_{t=t_a}^{t=t_b}\left[\frac{m}{2}\left(\frac{d}{dt}(\bar{x} + \eta)\right)^2 - \frac{1}{2}k(\bar{x} + \eta)^2\right]dt \quad\dots\dots\text{ Eq. (34).}$$

The idea is to expand S', given by the above expression, only upto terms linear in η and its derivatives, and then set S' equal to S_0 (given by Eq. (33)), by invoking the stationarity principle. Finally, obtain a differential equation for the least path $\bar{x}$. Expanding Eq. (34), we get:

Chapter 2: Variational Calculus

$$S' = \int_{t=t_a}^{t=t_b} \left[\frac{m}{2} \left(\frac{d}{dt}(\bar{x} + \eta) \right)^2 - \frac{1}{2} k(\bar{x} + \eta)^2 \right] dt$$

$$= \int_{t=t_a}^{t=t_b} \left[\frac{m}{2} \left(\frac{d\bar{x}}{dt} + \frac{d\eta}{dt} \right)^2 - \frac{1}{2} k(\bar{x} + \eta)^2 \right] dt$$

$$= \int_{t=t_a}^{t=t_b} \left[\frac{m}{2} \left(\left(\frac{d\bar{x}}{dt}\right)^2 + 2\frac{d\bar{x}}{dt}\frac{d\eta}{dt} + \left(\frac{d\eta}{dt}\right)^2 \right) - \frac{1}{2} k(\bar{x}^2 + 2\bar{x}\eta + \eta^2) \right] dt$$

[Using $(a + b)^2 = a^2 + 2ab + b^2$]

$$\approx \int_{t=t_a}^{t=t_b} \left[\left(\frac{m}{2} \left(\frac{d\bar{x}}{dt}\right)^2 + m \frac{d\bar{x}}{dt}\frac{d\eta}{dt} \right) - \frac{1}{2} k(\bar{x}^2 + 2\bar{x}\eta) \right] dt$$

[Neglecting the higher order terms $\left(\frac{d\eta}{dt}\right)^2$ and η^2]

$$= \int_{t=t_a}^{t=t_b} \left(\frac{m}{2} \left(\frac{d\bar{x}}{dt}\right)^2 - \frac{1}{2} k\bar{x}^2 \right) dt + \int_{t=t_a}^{t=t_b} \left(m \frac{d\bar{x}}{dt}\frac{d\eta}{dt} - k\bar{x}\eta \right) dt$$

[Regrouping the terms]

Note that the first integral in the above is S_0 (by Eq. (33)). Hence, we obtain from the above

$$S' = S_0 + \int_{t=t_a}^{t=t_b} \left[m \frac{d\bar{x}}{dt}\frac{d\eta}{dt} - k\bar{x}\eta \right] dt \quad \ldots\ldots\ldots\ldots\ldots \text{Eq. (35)}.$$

Integrating $\int_{t=t_a}^{t=t_b} \frac{d\bar{x}}{dt}\frac{d\eta}{dt} dt$ in Eq. (35) by parts, we obtain:

$$\int_{t=t_a}^{t=t_b} \frac{d\bar{x}}{dt}\frac{d\eta}{dt} dt = \frac{d\bar{x}}{dt} \eta(t) \Big|_{t=t_a}^{t=t_b} - \int_{t=t_a}^{t=t_b} \eta(t) \frac{d}{dt}\left(\frac{d\bar{x}}{dt}\right) dt$$

Now, since $\eta(t_b) = \eta(t_a) = 0$, the first term of the above equation vanishes, and we get:

$$\int_{t=t_a}^{t=t_b} \frac{d\bar{x}}{dt}\frac{d\eta}{dt} dt = - \int_{t=t_a}^{t=t_b} \eta(t) \frac{d}{dt}\left(\frac{d\bar{x}}{dt}\right) dt = - \int_{t=t_a}^{t=t_b} \eta(t) \frac{d^2\bar{x}}{dt^2} dt ,$$

using which in Eq. (35), we obtain:

$$S' = S_0 - m \int_{t=t_a}^{t=t_b} \eta(t) \frac{d^2\bar{x}}{dt^2} dt - \int_{t=t_a}^{t=t_b} k\bar{x}\eta \, dt$$

$$= S_0 - \int_{t=t_a}^{t=t_b} dt\,\eta(t) \left[m \frac{d^2\bar{x}}{dt^2} + k\bar{x} \right]$$

Since $S' = S_0$ per the stationary principle, we have from the above:

$$\int_{t=t_a}^{t=t_b} dt\,\eta(t) \left[m \frac{d^2\bar{x}}{dt^2} + k\bar{x} \right] = 0. \text{ As } \eta(t) \text{ is an arbitrary function, we get:}$$

Feynman's Path Integral explained with basic Calculus

$$m\frac{d^2\bar{x}}{dt^2} + k\bar{x} = 0 \quad \text{.. Eq. (36).}$$

Eq. (36) can be identified as Newton's law for a simple harmonic oscillator: you can see it more clearly by writing the equation as: $m\frac{d^2\bar{x}}{dt^2} = -k\bar{x}$, where $\frac{d^2\bar{x}}{dt^2}$ is the acceleration of the block of mass m and $-k\bar{x}$ is the force that the spring exerts on the block, when the spring is stretched by an amount $\bar{x}$. (See Figure 12.)

The solution to Eq. (36) is given by $\bar{x} = A\cos\omega t + B\sin\omega t$, where $\omega = \sqrt{\frac{k}{m}}$, and A and B are constants to be determined from the boundary conditions, viz., $\bar{x}(t = t_a) = x_a$ and $\bar{x}(t = t_b) = x_b$. Since $\bar{x}$ depends on sine and cosine, it's clearly oscillatory in nature.

Exercise 4. Check that $\bar{x} = A\cos\omega t + B\sin\omega t$, with $\omega = \sqrt{\frac{k}{m}}$, satisfies Eq. (36). Find A and B given that $\bar{x}(t = 0) = 0$ and $\bar{x}(t = t_b) = x_b$. Then evaluate S_0, the minimum value for the action S given by Eq. (32), i.e., calculate S for the least path or the classical path $\bar{x}(t)$ per Eq. (33): $S_0 = \int_{t=t_a}^{t=t_b} \left[\frac{1}{2}m\left(\frac{d\bar{x}}{dt}\right)^2 - \frac{1}{2}k\bar{x}^2\right] dt.$

Ans. $S_0 = \frac{m\omega x_b^2 \cos\omega t_b}{2\sin\omega t_b}$ .... Eq. (37).

We considered actions for three different scenarios: 1) free particle 2) a particle moving under the influence of gravity and 3) a particle connected to a spring. For the free particle, the potential energy PE is zero; for a particle moving in gravity, the $PE(= -mgy)$ is a linear function of the position (or space) variable; and for the spring mass system, the $PE\left(=\frac{1}{2}kx^2\right)$ is a quadratic function of the position (or space) variable. We obtained Newton's

Chapter 2: Variational Calculus

laws for these three systems by minimizing corresponding actions, following the Least Action Principle. We also calculated the actions for the respective "least" or "classical" paths of these three different systems.

RECAP: Least Action Principle:

It is a law of nature that a particle, moving under a force field, follows a path that minimizes the integral $S = \int_{t=t_a}^{t=t_b}[KE - PE]dt$, KE and PE being the kinetic and potential energies respectively. (S is the "action".) Hence, a particle's trajectory is the "least path" for the action, (i.e., the path that minimizes the action). The particle's trajectory is also called the "classical path". This fact of the classical path minimizing the corresponding action is called the Least Action Principle.

RECAP: The Actions Considered:

1. $S = \int_{t=t_a}^{t=t_b} \left[\frac{1}{2}m\left(\frac{dy}{dt}\right)^2\right] dt$ (Action for a free particle)

2. $S = \int_{t=t_a}^{t=t_b} \left[\frac{1}{2}m\left(\frac{dy}{dt}\right)^2 + mgy\right] dt$ (Action for a particle with a potential energy, which is linear in the position variable.)

3. $S = \int_{t=t_a}^{t=t_b} \left[\frac{1}{2}m\left(\frac{dx}{dt}\right)^2 - \frac{1}{2}kx^2\right] dt$ (Action for a particle with a potential energy that is quadratic in the position variable.)

Feynman's Path Integral explained with basic Calculus

Chapter 3: Feynman's Path Integral: a Mathematical Introduction

In this chapter we will talk about Feynman's Path Integral. We will first discuss the problem from a *purely mathematical point of view*. Once you are comfortable with the math, we will discuss the relevance of the Path Integral to quantum mechanics.

Discretizing an Action

As a first step to calculating Path Integrals, we need to discretize an action, meaning we will change the action's integral to a discrete sum. In the following, we discretize the action for a free particle $S = \int_{t=t_a}^{t=t_b} \frac{m}{2} \left(\frac{dx(t)}{dt} \right)^2 dt$, introduced in Chapter 2, Eq. (16)).[Unlike in Chapter 2, we label the position variable as x instead of y.]

In Figure 13, we divide the time interval from $t = t_a$ to $t = t_b$ into N equal parts, each of length $\epsilon = \frac{t_b - t_a}{N}$. [For this specific figure, we arbitrarily chose $N = 7$.] We label the time-points as $t_0, t_1, t_2, t_3, \ldots, t_{N-1}$ and t_N, the endpoints being $t_0 = t_a$ and $t_N = t_b$. The values of x at the two endpoints of the interval are: $x(t_0) =$

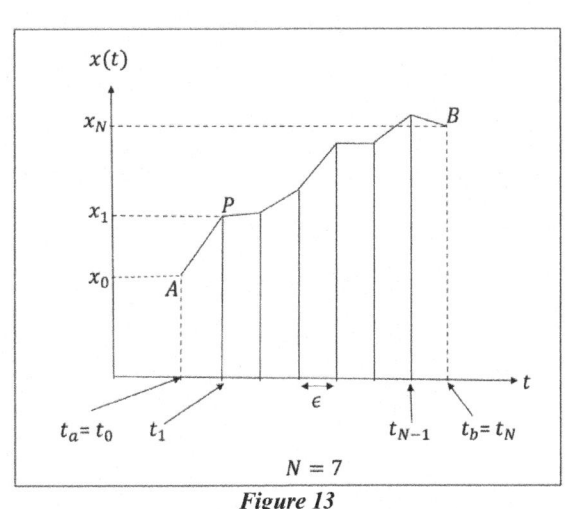

Figure 13

Feynman's Path Integral explained with basic Calculus

$x_0 \equiv x_a$ and $x(t_N) = x_N \equiv x_b$. The intermediate x-values at time-points t_1, $t_2, t_3, \ldots, t_{N-1}$ are denoted as $x_1, x_2, x_3, \ldots, x_{N-1}$ respectively. Hence, in the space-time graph, you have $N-1$ intermediate points $(t_1, x_1), (t_2, x_2), \ldots, (t_{N-1}, x_{N-1})$, and two endpoints A($t_a, x_a$) and B($t_b, x_b$). Joining these points consecutively with small line segments will produce a jagged-looking path, such as the one shown in Figure 13. Considering the first two consecutive points $A(t_0, x_0)$ and $P(t_1, x_1)$, the line-segment AP has the slope $\frac{dx}{dt} = \frac{x_1 - x_0}{\epsilon}$, which represents $\frac{dx}{dt}$ of the jagged path in time-interval between t_0 and t_1. Similarly, between t_1 and t_2, $\frac{dx}{dt}$ for the jagged path is: $\frac{dx}{dt} = \frac{x_2 - x_1}{\epsilon}$ and so on. Hence, the action $S = \int_{t=t_a}^{t=t_b} \frac{m}{2} \left(\frac{dx(t)}{dt}\right)^2 dt$ for the jagged path is given by:

$$S = \left(\frac{m}{2}\right)\left[\left(\frac{x_1 - x_0}{\epsilon}\right)^2 + \left(\frac{x_2 - x_1}{\epsilon}\right)^2 + \cdots + \left(\frac{x_{N-1} - x_{N-2}}{\epsilon}\right)^2 + \left(\frac{x_N - x_{N-1}}{\epsilon}\right)^2\right] \cdot \epsilon$$

.... Eq. (1).

As you can see, S, as given by Eq. (1), is a function of the "intermediate space variables": $x_1, x_2, x_3, \ldots, x_{N-1}$ (x_0 and x_N are fixed, since the endpoints A and B are fixed). Now, a jagged path such as the one in Figure 13, consists of one set of values for the variables $x_1, x_2, \ldots, x_{N-1}$. This set of values changes from one jagged path to another. Hence, the free-particle-action S, when written as a discrete sum as in Eq. (1), generally gives different values for different jagged paths having the same fixed endpoints.

In the next section, we will show how a discretized action is used to calculate a Path Integral.

Chapter 3: Feynman's Path Integral: a Mathematical Introduction

---

RECAP: Discretizing the Action for the Free Particle

Discretizing the free particle action $S = \int_{t=t_a}^{t=t_b} \left[\frac{1}{2}m\left(\frac{dx}{dt}\right)^2\right] dt$:

$$S = \frac{m}{2}\frac{(x_1-x_a)^2}{\epsilon} + \frac{m}{2}\frac{(x_2-x_1)^2}{\epsilon} + \cdots + \frac{m}{2}\frac{(x_{i+1}-x_i)^2}{\epsilon} + \cdots \frac{m}{2}\frac{(x_b-x_{N-1})^2}{\epsilon}$$

Calculating a Simple Path Integral

We explain the mathematics of the Path Integral with the following example. As you will see, an understanding of the discretization of S, discussed in the last section, will help you follow what is to come.

Let the time interval $t_b - t_a$ is divided into just two equal parts ($N = 2$), each of length $\epsilon = \frac{t_b - t_a}{2}$. (See Figure 14.) The time-points are t_0, t_1 and t_2, with $t_0 = t_a$ and $t_2 = t_b$ being the endpoints of the time-interval. $x(t_a) = x_a$ and $x(t_b) = x_b$. The x-variable at the intermediate time-point t_1 is x_1. Hence, the co-ordinates of the endpoints are $A \equiv (t_a, x_a)$ and $B \equiv (t_b, x_b)$, and the co-ordinates of the intermediate point are (t_1, x_1). We name the intermediate point C, which is not a fixed point, since for the same t_1, the space-variable x_1

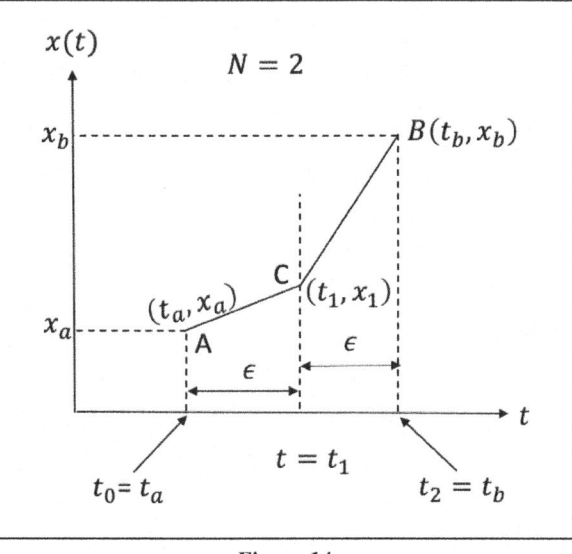

Figure 14

46

Feynman's Path Integral explained with basic Calculus

can assume different values. Hence, the jagged path ACB will be different for different values of x_1, i.e., different locations of C.

ACB consists of the segments AC and CB having different slopes. For AC, the slope $\frac{dx}{dt} = \frac{x_1-x_a}{\epsilon}$, whereas for CB, $\frac{dx}{dt} = \frac{x_b-x_1}{\epsilon}$. Hence, the free-particle action $S = \int_{t=t_a}^{t=t_b} \frac{1}{2}m \left(\frac{dx}{dt}\right)^2 dt$ is discretized as:

$$S = \left(\frac{1}{2}m\right)\left[\left(\frac{x_1-x_a}{\epsilon}\right)^2 + \left(\frac{x_b-x_1}{\epsilon}\right)^2\right]\epsilon$$

$$= \left(\frac{1}{2}m\right)\left[\frac{(x_1-x_a)^2}{\epsilon} + \frac{(x_b-x_1)^2}{\epsilon}\right] \quad \ldots\ldots\ldots\ldots\ldots \text{Eq. (2).}$$

As you can see from Eq. (2), S is a function of x_1 (x_a and x_b are constants). Now, consider the function e^{-kS}, where k is a positive constant and S is given by Eq. (2). Suppose you like to integrate e^{-kS} w.r.t x_1 just for fun. Denoting the integral as I, we get:

$$I = \int_{-\infty}^{+\infty} dx_1 \, e^{-kS}$$

$$= \int_{-\infty}^{+\infty} dx_1 \, e^{-\left(\frac{km}{2\epsilon}\right)[(x_1-x_a)^2+(x_b-x_1)^2]} \quad \ldots\ldots\ldots\ldots \text{Eq. (3).}$$

To evaluate the above integral, we will use the following formula we derived in the "Additional Exercises" in Chapter 1 (Eq. 16, Chapter 1) (reproduced in the following for convenience):

$$\int_{-\infty}^{+\infty} dx \, e^{[-k_1(x-a)^2 - k_2(x-b)^2]} = \sqrt{\frac{\pi}{k_1+k_2}} \, e^{\frac{-k_1 k_2}{k_1+k_2}(a-b)^2} \quad \ldots\ldots \text{Eq. (4).}$$

Using Eq. (4) with $k_1 = k_2 = \frac{km}{2\epsilon}$, we obtain from Eq. (3),

$$I = \int_{-\infty}^{+\infty} dx_1 \, e^{-kS}$$

$$= \sqrt{\frac{\pi\epsilon}{km}} \, e^{-\frac{km}{4\epsilon}(x_a-x_b)^2} \quad \ldots\ldots\ldots\ldots\ldots\ldots\ldots \text{Eq. (5).}$$

Note that in the above equation, I is a Gaussian function that depends on the difference of the x co-ordinates of the endpoints. Now, in Eq. (3), the integrand $e^{-\left(\frac{km}{2\epsilon}\right)[(x_1-x_a)^2+(x_b-x_1)^2]}$ is the product of the Gaussians $e^{-\left(\frac{km}{2\epsilon}\right)(x_1-x_a)^2}$ and

Chapter 3: Feynman's Path Integral: a Mathematical Introduction

$e^{-\left(\frac{km}{2\epsilon}\right)(x_1-x_b)^2}$ (which follows from the rule: $e^{p+q} = e^p e^q$). Hence, Eq. (5) shows that by integrating the product of two Gaussian functions or Gaussians $e^{-\left(\frac{km}{2\epsilon}\right)(x_1-x_a)^2}$ and $e^{-\left(\frac{km}{2\epsilon}\right)(x_1-x_b)^2}$ w.r.t to their common variable x_1, you get another Gaussian function.

How would Eq. (5) change, if we normalize each of the Gaussians we just mentioned, and replace those Gaussians in the integrand of Eq. (3) by their normalized versions? For normalizing the Gaussians $e^{-\left(\frac{km}{2\epsilon}\right)(x_1-x_a)^2}$ and $e^{-\left(\frac{km}{2\epsilon}\right)(x_1-x_b)^2}$, we need to multiply each by the factor:

$$\frac{1}{A} \equiv \sqrt{\frac{km}{2\pi\epsilon}} \quad \text{.. Eq. (6).}$$

(Refer to Eq. (3) of chapter 1, and the discussion leading up to Eq. (3) of Chapter 1 to bush up on normalization). Multiplication of the integrand in Eq. (3) (of the current chapter) by two $\frac{1}{A}$'s (one for each Gaussian), i.e., by a net factor of $\frac{1}{A} \cdot \frac{1}{A}$ will change the integral I in Eq. (5) to $I' = \left(\frac{1}{A}\right)^2 I$.

Hence,

$$I' = \left(\frac{1}{A}\right)^2 I$$

$$= \left(\frac{1}{A}\right)^2 \int_{-\infty}^{+\infty} dx_1 \, e^{-kS}$$

$$= \frac{km}{2\pi\epsilon} \sqrt{\frac{\pi\epsilon}{km}} \, e^{-\frac{km}{4\epsilon}(x_a-x_b)^2} \quad \text{[Using Eq. (5) and Eq. (6)]}$$

$$= \sqrt{\frac{km}{2\pi(2\epsilon)}} \, e^{-\frac{km}{2(2\epsilon)}(x_a-x_b)^2} \quad \text{.................................... Eq. (7).}$$

Note that 2ϵ appearing in Eq. (7) is simply $t_b - t_a$, since $\epsilon = \frac{t_b - t_a}{2}$, the time-interval $t_b - t_a$ having been divided into two equal parts.

48

Feynman's Path Integral explained with basic Calculus

What about carrying out a similar procedure, by dividing the interval $t_b - t_a$ into three equal parts ($N = 3$)? Then, each part will be of length $\epsilon = \frac{t_b - t_a}{3}$. In Figure 15, the time-points are t_0, t_1, t_2 and t_3, with $t_0 = t_a$ and $t_3 = t_b$ being the endpoints of the time-interval. $x(t_0) = x_a$ and $x(t_3) = x_b$. The x-values at the intermediate time-points t_1 and t_2 are x_1 and x_2 respectively. In the figure, the co-ordinates of the endpoints A and B are given by $A \equiv (t_a, x_a)$ and $B \equiv (t_b, x_b)$. The co-ordinates of the intermediate points (denoted by C and D respectively) are given by $C \equiv (t_1, x_1)$ and $D \equiv (t_2, x_2)$. Joining the points A, C, D, B by line-segments, we get the jagged path $ACDB$. Note that the jagged path is not a fixed one, since by changing x_1 and x_2 you can change the locations of C and D. Next, we will discretize the free-particle action $S = \int_{t=t_a}^{t=t_b} \frac{1}{2} m \left(\frac{dx}{dt}\right)^2 dt$ by using different values of $\frac{dx}{dt}$ for different line segments of the jagged path. For example, for AC, $\frac{dx}{dt} = \frac{x_1 - x_a}{\epsilon}$, whereas for CD, $\frac{dx}{dt} = \frac{x_2 - x_1}{\epsilon}$ and so on. Hence, the discretized S is given by:

$$S = \left(\frac{1}{2}m\right)\left[\left(\frac{x_1 - x_a}{\epsilon}\right)^2 + \left(\frac{x_2 - x_1}{\epsilon}\right)^2 + \left(\frac{x_b - x_2}{\epsilon}\right)^2\right]\epsilon$$

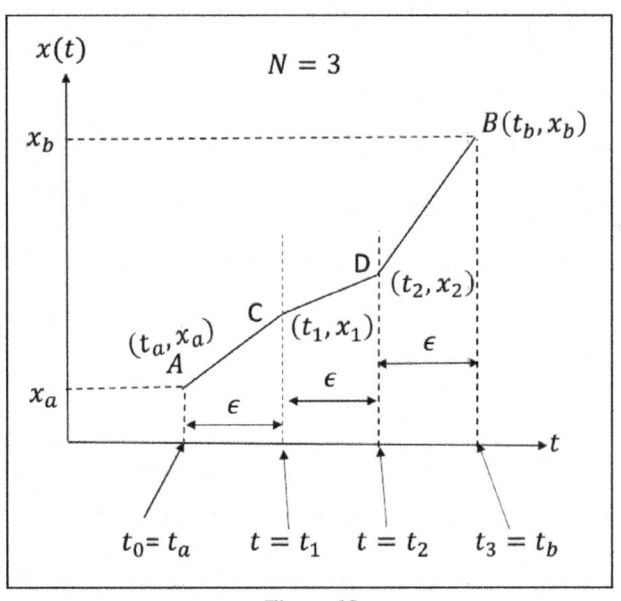

Figure 15

Chapter 3: Feynman's Path Integral: a Mathematical Introduction

$$= \left(\frac{1}{2}m\right)\left[\frac{(x_1-x_a)^2}{\epsilon} + \frac{(x_2-x_1)^2}{\epsilon} + \frac{(x_b-x_2)^2}{\epsilon}\right].$$

Using the above expression for S in e^{-kS}, we get, $e^{-kS} = e^{-\left(\frac{km}{2\epsilon}\right)[(x_1-x_a)^2+(x_2-x_1)^2+(x_b-x_2)^2]}$, which is a function of the variables x_1 and x_2, and also the product of "three" Gaussians $e^{-\left(\frac{km}{2\epsilon}\right)(x_1-x_a)^2}$, $e^{-\left(\frac{km}{2\epsilon}\right)(x_2-x_1)^2}$ and $e^{-\left(\frac{km}{2\epsilon}\right)(x_b-x_2)^2}$ (Using $e^{p+q+r} = e^p e^q e^r$ rule). To normalize the Gaussians, we need $\frac{1}{A} \equiv \sqrt{\frac{km}{2\pi\epsilon}}$ (same as given by Eq. (6)) for each of the Gaussians, resulting in $\left(\frac{1}{A}\right)^3$ multiplying e^{-kS}.[$\frac{1}{A}$ is multiplied three times, because there are "three" Gaussians.] We will next integrate $\left(\frac{1}{A}\right)^3 e^{-kS}$ along the lines similar to the $N = 2$ case, but since, this time, we have two intermediate x-variables, viz., x_1 and x_2, we will integrate twice as follows.

$$I = \left(\frac{1}{A}\right)^3 \int_{-\infty}^{+\infty}\int_{-\infty}^{+\infty} dx_1\, dx_2 e^{-kS}$$

$$= \left(\frac{1}{A}\right)^3 \int_{-\infty}^{+\infty}\int_{-\infty}^{+\infty} dx_1\, dx_2\, e^{-\left(\frac{km}{2\epsilon}\right)[(x_1-x_a)^2+(x_2-x_1)^2+(x_b-x_2)^2]} \quad \ldots \text{Eq. (8)}.$$

The integration variables x_1 and x_2 in Eq. (8) are the "intermediate" x-variables. Remember that there is no "reason" to either normalize the Gaussians or integrate with respect to the intermediate variables for either the $N = 2$ or the $N = 3$ case. Our activities are in the spirit of carrying out an experiment. Our steps for the current three-interval case are similar to those of the two-interval case: in both the cases, we normalize the Gaussians, then integrate w.r.t the intermediate x-variable(s). The goal of this experiment is to see if any general pattern emerges by studying the two and the three interval cases.

We will integrate Eq. (8) by first integrating w.r.t x_2, treating the other variable x_1 as constant, and then integrating w.r.t x_1. Hence, Eq. (8) is written as:

Feynman's Path Integral explained with basic Calculus

$$I = \left(\frac{1}{A}\right)^3 \int_{-\infty}^{+\infty} dx_1 e^{-\left(\frac{km}{2\epsilon}\right)[(x_1-x_a)^2]} \int_{-\infty}^{+\infty} dx_2\, e^{-\left(\frac{km}{2\epsilon}\right)[(x_2-x_1)^2+(x_b-x_2)^2]}$$

.... Eq. (9).

Carrying out the x_2 integral in Eq. (9) with the help of Eq. (4), we get:

$$\int_{-\infty}^{+\infty} dx_2\, e^{-\left(\frac{km}{2\epsilon}\right)[(x_2-x_1)^2+(x_b-x_2)^2]} = \sqrt{\frac{\pi\epsilon}{km}}\, e^{-\frac{km}{2(2\epsilon)}(x_b-x_1)^2},\text{ using}$$

which in Eq. (9), we get:

$$I = \left(\frac{1}{A}\right)^3 \sqrt{\frac{\pi\epsilon}{km}} \int_{-\infty}^{+\infty} dx_1 e^{-\left(\frac{km}{2\epsilon}\right)[(x_1-x_a)^2]} e^{-\frac{km}{2(2\epsilon)}(x_b-x_1)^2}.$$

Next, doing the x_1 integral in I (once again with the help of Eq. (4)), replacing $\frac{1}{A}$ by Eq. (6), and after some further algebra, we get:

$$I = \sqrt{\frac{km}{2\pi(3\epsilon)}}\, e^{-\frac{km}{2(3\epsilon)}(x_b-x_a)^2} \quad\quad\quad\quad\quad\quad\text{Eq. (10).}$$

Since, $3\epsilon = t_b - t_a$, Eq. (10) is similar to Eq. (7): If you replace 2ϵ appearing in Eq. (7), by 3ϵ, you get Eq. (10).

Now suppose you divide the time interval $[t_a, t_b]$ into N equal parts (i.e., $\epsilon = \frac{t_b - t_a}{N}$), meaning that the jagged path between A and B will have N line segments and $N - 1$ intermediate x-variables, viz., $x_1, x_2, \ldots, x_{N-1}$. Then the discretized action would be as given in Eq. (1) and reproduced in the following for convenience.

$$S = \frac{m}{2}\frac{(x_1-x_a)^2}{\epsilon} + \frac{m}{2}\frac{(x_2-x_1)^2}{\epsilon} + \cdots + \frac{m}{2}\frac{(x_{i+1}-x_i)^2}{\epsilon} + \cdots \frac{m}{2}\frac{(x_b-x_{N-1})^2}{\epsilon}$$

..............................Eq. (11).

$\frac{m}{2}\frac{(x_{i+1}-x_i)^2}{\epsilon}$ in the above expression for S is used to indicate a generic i_{th} term of the action S. i can be any number from 1 to $N - 1$.

Now, for the $N = 2$ case, we integrated $\left(\frac{1}{A}\right)^2 e^{-kS}$; for the $N = 3$ case, we integrated $\left(\frac{1}{A}\right)^3 e^{-kS}$. Can you guess what we would get if we integrated

51

Chapter 3: Feynman's Path Integral: a Mathematical Introduction

$\left(\frac{1}{A}\right)^N e^{-kS}$ with respect to the intermediate x-variables, viz., $x_1, x_2, \ldots, x_{N-1}$ as follows?

$I =$

$\left(\frac{1}{A}\right)^N \int_{-\infty}^{+\infty} dx_1 \int_{-\infty}^{+\infty} dx_2 \ldots \int_{-\infty}^{+\infty} dx_{N-2} \int_{-\infty}^{+\infty} dx_{N-1} \, e^{-kS}$ ………….Eq. (12),

where S is given by Eq. (11) and $\frac{1}{A}$ is given by Eq. (6)?

We answer that by studying the pattern: For the $N = 2$ case, we got: $I = \sqrt{\frac{km}{2\pi(2\epsilon)}} e^{-\frac{km}{2(2\epsilon)}(x_b - x_a)^2}$ (Eq. (7)), with the factor 2 in 2ϵ corresponding to $N = 2$; for the $N = 3$-interval case, we got: $I = \sqrt{\frac{km}{2\pi(3\epsilon)}} e^{-\frac{km}{2(3\epsilon)}(x_b - x_a)^2}$ (Eq. (10)), with 3ϵ corresponding to $N = 3$. Continuing this way, for the N-interval case, you will have $N\epsilon$ showing up in I at places where 2ϵ and 3ϵ previously showed up, producing the following expression for I.

$$I = \sqrt{\frac{km}{2\pi(N\epsilon)}} e^{-\frac{km}{2(N\epsilon)}(x_b - x_a)^2} \quad \text{Eq. (13).}$$

But since $\epsilon = \frac{t_b - t_a}{N}$, we have $N\epsilon = t_b - t_a$. Hence, from Eq. (13),

$$I = \sqrt{\frac{km}{2\pi(t_b - t_a)}} e^{-\frac{km}{2(t_b - t_a)}(x_b - x_a)^2} \quad \text{Eq. (14).}$$

Next we will let the number of divisions N go to infinity. (which means that the jagged path between the points A and B in the space-time diagram consists of "infinitely many" small line-segments.) Since $\epsilon = \frac{t_b - t_a}{N}$, $\epsilon \to 0$, as $N \to \infty$. But $N\epsilon$ in I (Eq. (13)) remains equal to $t_b - t_a$. Hence, the integral I (defined in Eq. (12)) produces Eq. (14) in the limit $N \to \infty$. Writing out explicitly (using only one integral sign to denote $N - 1$ integrals to keep the notation simple):

Feynman's Path Integral explained with basic Calculus

$$\lim_{N\to\infty} \left(\frac{1}{A}\right)^N \int_{-\infty}^{+\infty} dx_1 dx_2 \ldots dx_{N-1} \, e^{-\left(\frac{km}{2\epsilon}\right)[(x_1-x_a)^2+(x_2-x_1)^2+\cdots+(x_b-x_{N-1})^2]}$$

$$= \sqrt{\frac{km}{2\pi(t_b-t_a)}} \, e^{-\frac{km(x_b-x_a)^2}{2\,(t_b-t_a)}} \quad \ldots\ldots\ldots\ldots\ldots\ldots\ldots\ldots\ldots\ldots\ldots \text{Eq. (15)}.$$

Great job! Eq. (15) is the first Path Integral you did. You can think of the left of Eq. (15) as sum over paths, because a given set of values of $x_1, x_2, \ldots, x_{N-1}$ define a path, obtained by joining the points $(t_a, x_a), (t_1, x_1), (t_2, x_2), \ldots, (t_{N-1}, x_{N-1}), (t_b, x_b)$ consecutively (See Figure 13.). For that given path, the integrand $e^{-\left(\frac{km}{2\epsilon}\right)[(x_1-x_a)^2+(x_2-x_1)^2+\cdots+(x_{N-2}-x_{N-1})^2+(x_b-x_{N-1})^2]}$ assumes a specific value. When the variables $x_1, x_2, \ldots, x_{N-1}$ assume a different set of values, that means a different path and a different value for the integrand on the left of Eq. (15). So, as you vary the values of the variables $x_1, x_2, \ldots, x_{N-1}$, the integrand gets evaluated for different paths. Since a multiple integral is essentially summation over multiple variables, carrying out the integral in the left of Eq. (15) is equivalent to summing the integrand over different paths. That's how the path integral defined on the left of Eq. (15) assumes the interpretation of sum over paths. Using the standard notation of Path Integral, Eq. (15) is expressed as:

$$\int Dx[t] e^{-k \int_{t=t_a}^{t=t_b} \frac{1}{2} m \left(\frac{dx}{dt}\right)^2 dt} = \sqrt{\frac{km}{2\pi(t_b-t_a)}} \, e^{-\frac{km(x_b-x_a)^2}{2\,(t_b-t_a)}} \ldots\ldots \text{Eq. (16)}.$$

Note how $Dx[t]$ on the left side of Eq. (16) is a shorthand for $dx_1 dx_2 \ldots dx_{N-1}$ and the limiting process $N \to \infty$ combined. Also, note that although the action $\int_{t=t_a}^{t=t_b} \frac{m}{2} \left(\frac{dx(t)}{dt}\right)^2 dt$ on the left side of Eq. (16) is not written in a discretized form, it is implied that you would need to discretize the action to carry out the Path Integral. Basically, when you see an expression like the

Chapter 3: Feynman's Path Integral: a Mathematical Introduction

one on the left side of Eq. (16), you interpret it as what appears on the left side of Eq. (15).

We showed you the example of a Path Integral for a specific action (the free-particle-action). You can do Path Integrals for other actions by following a similar procedure involving discretizing the action. We will discuss such Path Integrals in later chapters in the context of quantum mechanics.

[Repeating the comment about the notation one more time to avoid confusion: The multiple integral $\int_{-\infty}^{+\infty} dx_1 dx_2 \ldots dx_{N-1}$ is the same as $\int_{-\infty}^{+\infty} dx_1 \int_{-\infty}^{+\infty} dx_2 \ldots \int_{-\infty}^{+\infty} dx_{N-2} \int_{-\infty}^{+\infty} dx_{N-1}$. Sometimes, we may also write the integration variables in the reverse order viz. $\int_{-\infty}^{+\infty} dx_{N-1} \ldots dx_2 dx_1$, also same as, $\int_{-\infty}^{+\infty} dx_{N-1} \int_{-\infty}^{+\infty} dx_{N-2} \ldots \int_{-\infty}^{+\infty} dx_2 \int_{-\infty}^{+\infty} dx_1$. All of these variations mean the same.]

Role of the Constant

Let us talk about ordinary integration for a moment. How do you calculate $\int_a^b f(x)dx$? You divide the x-interval (a, b) into a very large number of small intervals, each of width ϵ. Corresponding to each small interval, you construct a narrow rectangle, whose height is equal to the value of the function at the endpoint of that interval (See

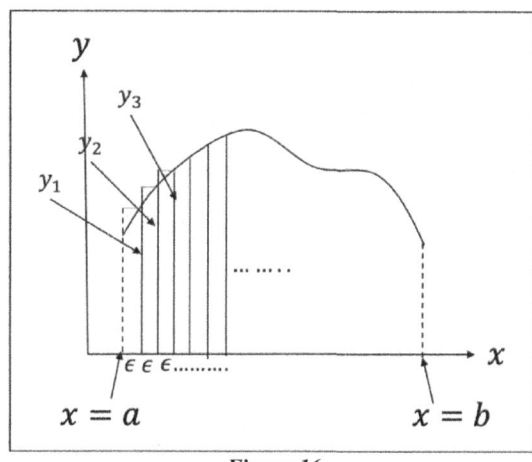

Figure 16

Feynman's Path Integral explained with basic Calculus

Figure 16). For example, the area of the first rectangle is: ϵy_1. Similarly, the area of the second rectangle is: ϵy_2 and so on. For small ϵ, the sum of the areas of the rectangles closely follows the area under the curve as shown in the diagram. Writing the area under the curve as $\int_a^b f(x)dx$:

$$\int_a^b f(x)dx \approx \epsilon(y_1 + y_2 + y_3 + \cdots) \quad \text{................ Eq. (17).}$$

Now on the right side of Eq. (17), when you let the width ϵ go to zero, the sum of y-vaules tend to infinity, because you have an infinite number of the y's in the limit $\epsilon \to 0$. But, as soon as you multiply that sum $(y_1 + y_2 + y_3 + \cdots)$ by the small ϵ, the product starts approaching a finite value. This finite value $\epsilon(y_1 + y_2 + y_3 + \cdots.)$, (with $\epsilon \to 0$) is the area under the graph of y between the vertical lines $x = a$ and $x = b$.

So, ordinary integration is not just summation, you need to multiply the sum of the y-values by a normalizing constant ϵ to obtain a finite limiting value. In a similar vein, in the Path Integral, $\left(\frac{1}{A}\right)^N$ plays the role of the normalizing constant. By multiplying the multiple integral on the left of Eq. (15) by $\left(\frac{1}{A}\right)^N$, you get a non-trivial, finite result as you take the limit $N \to \infty$.

Exercise 1. Assume that the time interval $t_b - t_a$ is divided into four equal parts ($N = 4$), each of length ϵ. The endpoints of the time-interval are $t_0 = t_a$ and $t_4 = t_b$ with $x(t_0) = x_a$ and $x(t_4) = x_b$. The intermediate time-points are $t_1, t_2,$ and t_3. The intermediate x-values at time-points $t_1, t_2,$ and t_3 are denoted as $x_1, x_2,$ and x_3 respectively.

The discretized free-particle-action can be written follows.

$$S = \left(\frac{1}{2}m\right)\left[\left(\frac{x_1-x_a}{\epsilon}\right)^2 + \left(\frac{x_2-x_1}{\epsilon}\right)^2 + \left(\frac{x_3-x_2}{\epsilon}\right)^2 + \left(\frac{x_b-x_3}{\epsilon}\right)^2\right]\epsilon$$

$$= \left(\frac{1}{2}m\right)\left[\frac{(x_1-x_a)^2}{\epsilon} + \frac{(x_2-x_1)^2}{\epsilon} + \frac{(x_3-x_2)^2}{\epsilon} + \frac{(x_b-x_3)^2}{\epsilon}\right]$$

Chapter 3: Feynman's Path Integral: a Mathematical Introduction

Carry out the following iterated multiple integral:

$$I = \left(\frac{1}{A}\right)^4 \int_{-\infty}^{+\infty} dx_1 \int_{-\infty}^{+\infty} dx_2 \int_{-\infty}^{+\infty} dx_3 \, e^{-kS}$$

$$= \left(\frac{1}{A}\right)^4 \int_{-\infty}^{+\infty} dx_1 \int_{-\infty}^{+\infty} dx_2 \int_{-\infty}^{+\infty} dx_3 \, e^{-\frac{km}{2}\left[\frac{(x_1-x_a)^2}{\epsilon} + \frac{(x_2-x_1)^2}{\epsilon} + \frac{(x_3-x_2)^2}{\epsilon} + \frac{(x_b-x_3)^2}{\epsilon}\right]},$$

where $\frac{1}{A} \equiv \sqrt{\frac{km}{2\pi\epsilon}}$ (Eq. (6)). You will need to repeatedly use Eq. (4). Check that your final result is formally similar to the right side of Eq. (15).

Indirect Method for Evaluating Path Integrals

In this section, you will learn how to evaluate the same Path Integral you did in the previous section by an indirect method. This method involves less work, since you will not need to do the Gaussian integrals. But this method allows you to ascertain the Path Integral's value only up to a constant. That's the price you pay for carrying out your calculations indirectly and easily. However, in physical problems, a lot can be learned without needing to know the multiplicative constant; besides, the constants can be determined from other considerations as you will see later in the book. The indirect method works for certain actions such as the free-particle action $S = \int_{t=t_a}^{t=t_b} \frac{1}{2} m \left(\frac{dx}{dt}\right)^2 dt$ and others that you will come across later. You have already evaluated the Path Integral for $S = \int_{t=t_a}^{t=t_b} \frac{1}{2} m \left(\frac{dx}{dt}\right)^2 dt$ directly in the last section. In the following, we evaluate the Path Integral for the same action by using the indirect method. As before, the Path Integral is given by:

$$I = \lim_{N \to \infty} \left(\frac{1}{A}\right)^N \int_{-\infty}^{+\infty} dx_{N-1} \int_{-\infty}^{+\infty} dx_{N-2} \ldots \ldots \int_{-\infty}^{+\infty} dx_3 \int_{-\infty}^{+\infty} dx_2 \int_{-\infty}^{+\infty} dx_1 \, e^{-kS}$$

..................Eq. (18).

Feynman's Path Integral explained with basic Calculus

The idea behind the indirect method is to write Eq. (18) in terms of a new set of variables.

In Figure 17, let $\bar{x}(t)$ be the "least path" for S defined between two given endpoints A and B. [See Chapter 2 to brush up on the least path. Also, we proved in chapter 2 that the least path for the free-particle action S is a straight line joining the points A and B. Nevertheless, in Figure 17, the least path looks arbitrary, which is intentional, because there are situations where a least path isn't a straight line, and yet the "indirect method" for evaluating the Path Integral works. Hence, to emphasize the broader applicability of the indirect method, we assumed an arbitrary shape for the least path $\bar{x}(t)$ in Figure 17.] By adding an arbitrary variation $\eta(t)$ (not necessarily small) to the least path $\bar{x}(t)$, we obtain an arbitrary path $x(t)$ between the endpoints A and B, given by: $x(t) = \bar{x}(t) + \eta(t)$. The jagged path in Figure 17 represents the arbitrary path $x(t)$. Since $x(t)$ and $\bar{x}(t)$ have the same values at $t = t_a$ and $t = t_b$, $\eta(t)$ must vanish at $t = t_a$ and $t = t_b$, i.e., $\eta(t_a) = \eta(t_b) = 0$.

Now, $x_1 \equiv x(t = t_1)$
$= \bar{x}(t = t_1) + \eta(t = t_1)$ (Using $x(t) = \bar{x}(t) + \eta(t)$)
$\equiv \bar{x}_1 + \eta_1$ (We used the definitions: $\bar{x}(t = t_1) \equiv \bar{x}_1$
and $\eta(t = t_1) \equiv \eta_1$)

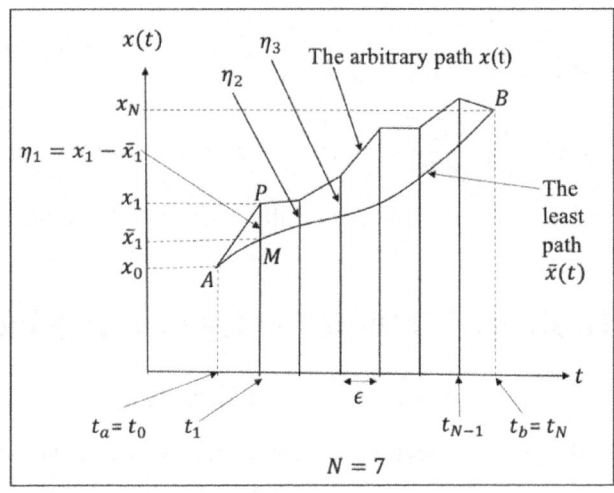

Figure 17

Chapter 3: Feynman's Path Integral: a Mathematical Introduction

So, $x_1 = \bar{x}_1 + \eta_1$. Similarly, $x_2 \equiv x(t = t_2) = \bar{x}_2 + \eta_2$,.....and so on. So, we have a new set of variables $\eta_1, \eta_2, \ldots \eta_{N-1}$ corresponding to $N-1$ intermediate variables $x_1, x_2, \ldots x_{N-1}$.

First, considering $x_1 = \bar{x}_1 + \eta_1$, we have, $dx_1 = d\bar{x}_1 + d\eta_1$. But $d\bar{x}_1 = 0$, since $\bar{x}_1$, being on the least path, is a constant (the least path is fixed). So, $dx_1 = d\eta_1$. Similarly, $dx_2 = d\eta_2$, and so on. We write Eq. (18) in terms of the new variables η's as follows:

$$I = \lim_{N \to \infty} \left(\frac{1}{A}\right)^N \int_{-\infty}^{+\infty} d\eta_{N-1} \int_{-\infty}^{+\infty} d\eta_{N-2} \ldots \ldots \int_{-\infty}^{+\infty} d\eta_3 \int_{-\infty}^{+\infty} d\eta_2 \int_{-\infty}^{+\infty} d\eta_1 \, e^{-kS}$$

.... Eq. (19),

We will evaluate Eq. (19) for the free-particle action $S = \int_{t=t_a}^{t=t_b} \frac{1}{2} m \left(\frac{dx}{dt}\right)^2 dt$ by writing $x = \bar{x} + \eta$ and expanding S about $\bar{x}$ as follows: [You will find the following calculation partly similar to that of Example 1 (of Chapter 2).]

$$S = \frac{m}{2} \int_{t=t_a}^{t=t_b} \left(\frac{d}{dt}(\bar{x} + \eta)\right)^2 dt \quad [\text{Using } x = \bar{x} + \eta]$$

$$= \frac{m}{2} \int_{t=t_a}^{t=t_b} \left(\frac{d}{dt}(\bar{x} + \eta)\right)^2 dt$$

$$= \frac{m}{2} \int_{t=t_a}^{t=t_b} \left(\frac{d\bar{x}}{dt} + \frac{d\eta}{dt}\right)^2 dt$$

Expanding the above integrand using $(a + b)^2 = a^2 + 2ab + b^2$, we get:

$$S = \frac{m}{2} \int_{t=t_a}^{t=t_b} \left(\frac{d\bar{x}}{dt}\right)^2 dt + m \int_{t=t_a}^{t=t_b} \frac{d\bar{x}}{dt} \frac{d\eta}{dt} dt + \frac{m}{2} \int_{t=t_a}^{t=t_b} \left(\frac{d\eta}{dt}\right)^2$$

.... Eq. (20).

In the above integral, we will carry out integration by parts on the second integral: $\int_{t=t_a}^{t=t_b} \frac{d\bar{x}}{dt} \frac{d\eta}{dt} dt$. We assume $u = \frac{d\bar{x}}{dt}$, and $dv = \frac{d\eta}{dt} dt$. Using $\int u \, dv = uv - \int v \, du$, we obtain:

$$\int_{t=t_a}^{t=t_b} \frac{d\bar{x}}{dt} \frac{d\eta}{dt} dt = \frac{d\bar{x}}{dt} \eta(t) \Big|_{t=t_a}^{t=t_b} - \int_{t=t_a}^{t=t_b} \eta(t) \frac{d}{dt}\left(\frac{d\bar{x}}{dt}\right) dt$$

Feynman's Path Integral explained with basic Calculus

Now, since $\eta(t = t_a) = \eta(t = t_b) = 0$, the first term on the right of the above equation vanishes, and we are left with the second term. Hence, $\int_{t=t_a}^{t=t_b} \frac{d\bar{x}}{dt} \frac{d\eta}{dt} dt = -\int_{t=t_a}^{t=t_b} \eta(t) \frac{d^2\bar{x}}{dt^2} dt$, using which in Eq. (20), we obtain:

$$S = \frac{m}{2} \int_{t=t_a}^{t=t_b} \left[\left(\frac{d\bar{x}}{dt}\right)^2\right] dt - m \int_{t=t_a}^{t=t_b} \eta(t) \frac{d^2\bar{x}}{dt^2} dt + \frac{m}{2} \int_{t=t_a}^{t=t_b} \left(\frac{d\eta}{dt}\right)^2 dt \dots \text{Eq. (21)}.$$

The first integral in Eq. (21) is simply the free particle action $S = \int_{t=t_a}^{t=t_b} \frac{1}{2} m \left(\frac{dx}{dt}\right)^2 dt$ evaluated for the least path $x(t) = \bar{x}(t)$. Referring to $\frac{m}{2} \int_{t=t_a}^{t=t_b} \left[\left(\frac{d\bar{x}}{dt}\right)^2\right] dt$ as S_0, we obtain from Eq. (21):

$$S = S_0 - m \int_{t=t_a}^{t=t_b} \eta(t) \frac{d^2\bar{x}}{dt^2} dt + \frac{m}{2} \int_{t=t_a}^{t=t_b} \left(\frac{d\eta}{dt}\right)^2 dt \dots \text{Eq. (22)}.$$

Applying the stationarity principle (introduced in Chapter 2), S is equal to S_0 (the action for the least path) up to the first order variation in η. Hence, in Eq. (22), ignoring the third term (which is the square of the derivative of η an hence not "first order" or "linear" in η) for the moment and setting S equal to S_0, we get $\int_{t=t_a}^{t=t_b} \eta(t) \frac{d^2\bar{x}}{dt^2} dt = 0$, which implies $\frac{d^2\bar{x}}{dt^2} = 0$, since $\eta(t)$ is arbitrary. Plugging $\frac{d^2\bar{x}}{dt^2} = 0$ back in Eq. (22), we get:

$$S = S_0 + \frac{m}{2} \int_{t=t_a}^{t=t_b} \left(\frac{d\eta}{dt}\right)^2 dt \dots \text{Eq. (23)}.$$

Ignoring the higher order term is nothing new. We did that in Chapter 2, to obtain $\frac{d^2\bar{x}}{dt^2} = 0$ (Eq. (22) of Chapter 2; there we used a different variable name.) In this case, we ignore the higher order term again, but only momentarily, to apply *the stationarity principle*. The higher order term is very much present in the action S, as is clear from Eq. (23) of the current chapter. Since S is expanded about the least path, S doesn't have a linear term in η (which represents a variation about the least path) in Eq. (23). This is analogous to the idea that if you write the equation of a parabola by using its

Chapter 3: Feynman's Path Integral: a Mathematical Introduction

minimum as the origin, then you will not have a linear term in the parabola's equation anymore (See Figure 18). As shown in the diagram, when the parabola's vertex (the minimum point) doesn't coincide with the

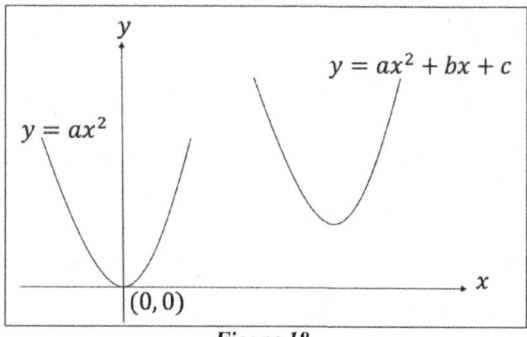

Figure 18

origin, the equation of the parabola contains a linear term in x, which is absent when the parabola's minimum (vertex) coincides with the origin.

Notice how S_0, the integral involving $\bar{x}(t)$, is completely separated from an integral involving $\eta(t)$ in Eq. (23). Denoting the second term in Eq. (23) as S_η (the suffix of S conveys that the integral depends on the function $\eta(t)$), we have:

$$S = S_0 + S_\eta, \text{ where } S_\eta = \frac{m}{2}\int_{t=t_a}^{t=t_b} \left(\frac{d\eta}{dt}\right)^2 dt$$

$$\Rightarrow e^{-kS} = e^{-k(S_0+S_\eta)} = e^{-kS_0}e^{-kS_\eta}$$, using which in the Path Integral of Eq. (19), we obtain:

$$I = \lim_{N\to\infty} \left(\frac{1}{A}\right)^N \int_{-\infty}^{+\infty} d\eta_{N-1} \int_{-\infty}^{+\infty} d\eta_{N-2} \ldots \ldots \int_{-\infty}^{+\infty} d\eta_2 \int_{-\infty}^{+\infty} d\eta_1 \, e^{-kS_0}e^{-kS_\eta}$$

Since, e^{-kS_0} in the above integrand does not depend on η's, we can take it outside the integral. Hence:

$$I = e^{-kS_0} \lim_{N\to\infty} \left(\frac{1}{A}\right)^N \int_{-\infty}^{+\infty} d\eta_{N-1} \int_{-\infty}^{+\infty} d\eta_{N-2} \ldots \ldots \int_{-\infty}^{+\infty} d\eta_2 \int_{-\infty}^{+\infty} d\eta_1 \, e^{-kS_\eta}$$

.... Eq. (24).

Now, the multiple integral $\lim_{N\to\infty} \left(\frac{1}{A}\right)^N \int_{-\infty}^{+\infty} d\eta_{N-1} \ldots \int_{-\infty}^{+\infty} d\eta_1 \, e^{-kS_\eta}$, in Eq. (24), with $S_\eta = \frac{m}{2}\int_{t=t_a}^{t=t_b}\left(\frac{d\eta}{dt}\right)^2 dt$ can be written using the previously introduced notation of Path integrals as $\int D\eta[t]e^{-kS_\eta} = $

Feynman's Path Integral explained with basic Calculus

$\int D\eta[t] e^{-k \int_{t=t_a}^{t=t_b} \frac{1}{2} m \left(\frac{d\eta}{dt} \right)^2 dt}$, where $D\eta[t]$ serves as a shorthand for $\lim_{N \to \infty} \left(\frac{1}{A} \right)^N \int_{-\infty}^{+\infty} d\eta_{N-1} \ldots \int_{-\infty}^{+\infty} d\eta_1$. The Path Integral involves only η-variables and looks exactly like the left side of Eq. (16) except for a different variable name. Since $\eta(t_a) = \eta(t_b) = 0$, we could evaluate the η-Path Integral by setting the boundary conditions $x_a = x_b = 0$ on the right of Eq. (16). (See the Example below.) But in an indirect method, we do not bother evaluating a Path Integral that depends only on η's. Instead, we treat the η-Path Integral as a constant factor that multiplies the exponential e^{-kS_0} (see Eq. (24)). (Note that for this specific problem, the Path Integral involving the η-variables could be evaluated directly (if one wished) with relative ease, but there are cases, as in an example in the next chapter, when doing such an integral involving the η variables is hard). We write Eq. (24) as,

$$I = e^{-kS_0}(C) \ldots\ldots\ldots\ldots\ldots\ldots\ldots\ldots\ldots\ldots\ldots\ldots \text{Eq. (25)}.$$

where the constant $C =$ is given by:

$$C = \int D\eta[t] e^{-kS_\eta} = \int D\eta[t] e^{-k \int_{t=t_a}^{t=t_b} \frac{1}{2} m \left(\frac{d\eta}{dt} \right)^2 dt} \ldots\ldots\ldots \text{Eq. (26)}.$$

We calculated S_0, the free-particle-action evaluated for the corresponding least path, in Eq. (23) of Chapter 2. Using that S_0 now in Eq. (25) we get:

$$I = e^{-\frac{km(x_b - x_a)^2}{2(t_b - t_a)}} C \ldots\ldots\ldots\ldots\ldots\ldots\ldots\ldots\ldots\ldots \text{Eq. (27)}.$$

Notice the partial agreement between Eq. (27), in which the Path Integral is calculated using the indirect method and Eq. (16), in which the same Path Integral is evaluated directly: Eq. (27) produces the exponential factor of Eq. (16) but cannot determine Eq. (16) wholly.

Example. Calculate C, as given by Eq. (26). Then, using Eq. (26) in Eq. (27), check that Eq. (27) gives the same result as Eq. (16).

Chapter 3: Feynman's Path Integral: a Mathematical Introduction

<u>Soln.</u> The Path Integral, given by C in Eq. (26), is formally the same as what you have on the left side of Eq. (16): you only have to relabel x in Eq. (16) as η. However, in Eq. (26), η satisfies the boundary conditions $\eta(t_a) = \eta(t_b) = 0$; so we should set $x_a = x_b = 0$ in Eq. (16), to get C in Eq. (26). Setting $x_a = x_b = 0$ on the right of Eq. (16), we get $\sqrt{\frac{km}{2\pi(t_b-t_a)}}$; hence, $C = \sqrt{\frac{km}{2\pi(t_b-t_a)}}$, using which in Eq. (27), we obtain $I = e^{-\frac{km(x_b-x_a)^2}{2\,(t_b-t_a)}} \sqrt{\frac{km}{2\pi(t_b-t_a)}}$, same as what we have in Eq. (16).

At this point, I have two choices: One, I can show you more examples of Path Integrals involving e^{-kS} type of integrand for different actions S. But that will take us away from quantum mechanics, since in quantum mechanics, Path Integrals involve exponentials with imaginary exponents (e^{ikS}). My second choice is to first explain why Path Integrals show up in quantum mechanics, then show you more examples of the Path Integral involving the integrand e^{ikS}. I will take the second approach, since A) our objective is to discuss Path Integrals in the context of quantum mechanics, and B) whether you use e^{-kS} or e^{ikS}, the technique of carrying out Path Integrals is essentially the same.

Feynman's Path Integral explained with basic Calculus

Chapter 4: Path Integrals in Quantum Mechanics

The Connection between Classical and Quantum

In classical mechanics, an object, obeying Newton's law in a given force field, follows a definitive path. On the other hand, quantum mechanics is inherently probabilistic and does not entertain the concept of particle trajectories. In quantum mechanics, we speak the language of probability amplitudes, the absolute square of which give probabilities (more accurately, probability densities). For example, imagine a non-relativistic particle in a force field. Starting at A (in the space-time diagram of Figure 19), the particle has a finite *probability amplitude* to reach the point B. Such a probability amplitude, denoted by $K(B,A)$, is also called a *propagator*. (Note how in the notation of the propagator, the "final" point B comes before the "initial" point A). $|K(B,A)|^2$ gives the probability distribution function for the particle to reach B after starting at A.

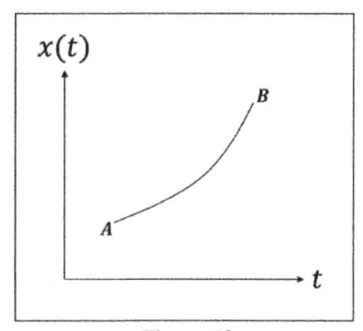

Figure 19

In the following, we will explain how a propagator $K(B,A)$ is calculated. You will also see how the "classical path" – the actual path that the particle would have taken had it moved following Newton's law – plays a prominent role in the quantum world as well, the propagator acting as the bridge between the quantum and classical worlds. To understand how, we need to first understand how a propagator $K(B,A)$ is calculated.

Feynman's Path Integral explained with basic Calculus

In Figure 20, let the path marked 0 denote the classical path of a particle moving under some force field. There are various other paths joining A and B. For each of the paths (whether classical or not), you can calculate an action: $S = \int_{t_a}^{t_b}(KE - PE)dt$. For example, the action for the classical path (labeled 0 in the diagram) is S_0; the action for the path "1" is S_1 and so on. Per the Least Action principle (introduced in Chapter 2), the classical path minimizes the action S and hence is the least path.

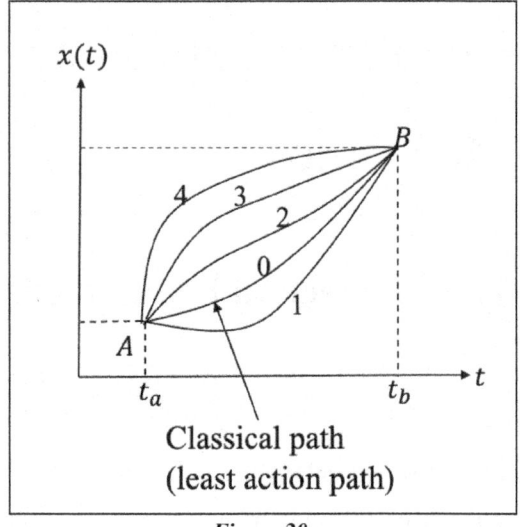

Figure 20 — Classical path (least action path)

Feynman postulated that the propagator $K(B, A)$ can be calculated by adding up complex numbers $e^{\frac{i}{\hbar}S}$'s for all possible paths between the points A and B ($\hbar \equiv \frac{h}{2\pi}$, where h is the Planck's constant; and i is the imaginary number.). So, if we number the paths as 0, 1, 2, 3,....etc., the propagator, per Feynman, is proportional to the sum of the corresponding complex exponentials, as given by the following:

$$K(B, A) \sim e^{\frac{i}{\hbar}S_0} + e^{\frac{i}{\hbar}S_1} + e^{\frac{i}{\hbar}S_2} + e^{\frac{i}{\hbar}S_3} + e^{\frac{i}{\hbar}S_4} + \cdots \quad \text{Eq. (1)}.$$

Since Eq. (1) involves sum over paths, we can and will rigorously define the sum on the right of Eq. (1) as a Path Integral. But, for now, let us analyze the sum qualitatively, by drawing vectors or arrows joined end to end. Why arrows? Because complex numbers are represented by directed line-segments. In Figure 21, a vector of length C, starting at the origin O, and making an angle ϕ with the x-axis, represents the complex number $Ce^{i\phi}$. To add another

Chapter 4: Path Integrals in Quantum Mechanics

complex number $De^{i\theta}$ to it, we simply draw a second vector (representing $De^{i\theta}$) from the tip of the first one (Q), as shown in Figure 21. The tip of the second vector is P. The sum $Ce^{i\phi} + De^{i\theta}$ is represented by the arrow (vector) joining O to the final point P.

Similarly, in Eq. (1), the complex numbers $e^{\frac{i}{\hbar}S_0}$, $e^{\frac{i}{\hbar}S_1}$, $e^{\frac{i}{\hbar}S_2}$, ...etc., are

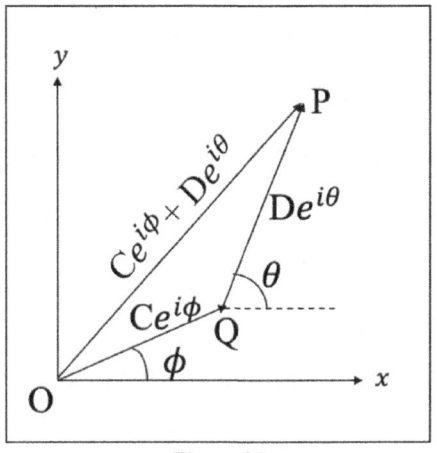

Figure 21

all *arrows* of unit length, making angles $\frac{S_0}{\hbar}, \frac{S_1}{\hbar}, \frac{S_2}{\hbar}$,.. etc with the x-axis and their sum is carried out by joining the arrows end to end. But if the angles are wildly different from each other, the final point may not be very far away from the initial, (as shown in Figure 22(a)), implying the sum given by Eq. (1) is small. But if the angles $\frac{S_0}{\hbar}, \frac{S_1}{\hbar}, \frac{S_2}{\hbar}$,.. etc are equal to each other for whatever reasons, i.e., if the actions $S_0, S_1, S_2, S_3, \ldots$ are the same, the arrows are aligned (as shown in Figure 22(b), and the vector joining the beginning and the end points, has a substantial magnitude. For what sort of paths $0, 1, 2, 3, \ldots$etc., will the corresponding actions $S_0, S_1, S_2, S_3, \ldots$ be possibly equal or at least close to each other? Answer: If the paths 1, 2, 3, ... vary only "slightly" from the classical path 0 (also the least path). Because, per the stationarity principle discussed in chapter 2, for slight variations about the least path, the action stays

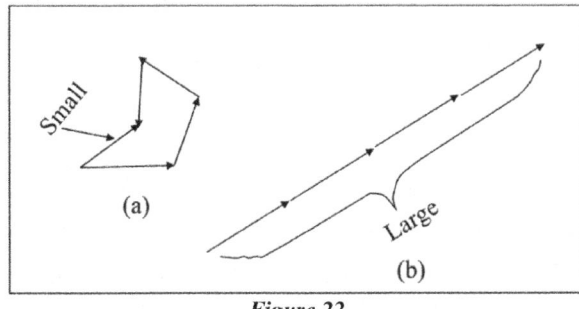

Figure 22

Feynman's Path Integral explained with basic Calculus

the same (up to the first order of the variations). In other words, the classical path and those differing only slightly from it contribute most to the calculation of the propagator $K(B,A)$ in Eq. (1)! This is how the classical path (the actual trajectory) continues to play an important role even in quantum mechanics. "Summing $e^{\frac{i}{\hbar}S}$ over all paths" brings out this remarkable connection between the classical and quantum.

Formal Definition of a Propagator

Next, we will define $K(B,A)$, introduced in Eq. (1), precisely, by means of the Path Integral. First of all, instead of using the notation $K(B,A)$, we will use the alternative notation $K(t_b, x_b; t_a, x_a)$, where (t_a, x_a) and (t_b, x_b) are the co-ordinates of the initial point A and the final point B respectively. Drawing inspiration from the only other Path Integral we have seen so far, i.e., Eq. (12) of Chapter 3 (with $N \to \infty$), we formally define summing $e^{\frac{i}{\hbar}S}$ over paths as the following Path Integral.

$$K(t_b, x_b; t_a, x_a)$$
$$= \lim_{N \to \infty} \left(\frac{1}{A}\right)^N \int_{-\infty}^{+\infty} dx_{N-1} \int_{-\infty}^{+\infty} dx_{N-2} \ldots \ldots \int_{-\infty}^{+\infty} dx_3 \int_{-\infty}^{+\infty} dx_2 \int_{-\infty}^{+\infty} dx_1 \, e^{\frac{i}{\hbar}S}$$
.... Eq. (2).

In Eq. (2), you will use different S's for different physical systems. For example, while calculating the propagator for a particle moving freely, i.e., in the absence of any force, we should use the free-particle-action for S. For calculating the propagator of a quantum particle having a quadratic potential energy, we will use S with a quadratic potential energy (like the spring mass system) and so on. What about the factor $\frac{1}{A}$ in Eq. (2)? How to find it? Will they be different for different physical systems? In Chapter 6, you will learn how to determine $\frac{1}{A}$ in course of deriving Schrodinger's equation using the

Chapter 4: Path Integrals in Quantum Mechanics

propagator. You will also see how you get the same $\frac{1}{A}$ regardless of the potential energies used in the action. However, it is possible to guess $\frac{1}{A}$ correctly for the "free-particle case", as we will show in our next section by drawing analogy with the Path Integral we calculated in Chapter 3.

Free Particle Propagator

In this section, we assume the argument of the complex exponential in Eq. (2) is the free-particle-action S (discretized), given by Eq. (11) of Chapter 3, and reproduced in the following for convenience:

$$S = \left(\frac{m}{2}\right)\left[\frac{(x_1-x_a)^2}{\epsilon} + \frac{(x_2-x_1)^2}{\epsilon} + \cdots + \frac{(x_{N-1}-x_{N-2})^2}{\epsilon} + \frac{(x_b-x_{N-1})^2}{\epsilon}\right]$$

..........................Eq. (3).

Using Eq. (3) in Eq. (2), we obtain:

$$K(t_b, x_b; t_a, x_a)$$
$$= \lim_{N\to\infty} \left(\frac{1}{A}\right)^N \int_{-\infty}^{+\infty} dx_1 dx_2 \ldots dx_{N-1}\, e^{\left(\frac{im}{2\hbar\epsilon}\right)[(x_1-x_a)^2 + (x_2-x_1)^2 + \cdots + (x_b-x_{N-1})^2]}$$

.... Eq. (4).

Note the similarity between Eq. (4) and what you have on the left side of Eq. (15) of Chapter 3: $\frac{i}{\hbar}$ in Eq. (4) (of the current chapter) is equivalent to $-k$ on the left side of Eq. (15) of Chapter 3.

So, $\frac{i}{\hbar} \Leftrightarrow -k$, implying $k \Leftrightarrow -\frac{i}{\hbar}$. Exploiting this equivalence, we carry out the following two steps.

1. We replace k by $-\frac{i}{\hbar}$ in the expression of $\frac{1}{A}$ given in Eq. (6) of Chapter 3 to get $\frac{1}{A}$ for our current problem. So,

$$\frac{1}{A} = \sqrt{\frac{km}{2\pi\epsilon}} = \sqrt{\frac{-im}{\hbar 2\pi\epsilon}} = \sqrt{\frac{m}{2\pi i\hbar\epsilon}}$$ (It is this value of $\frac{1}{A}$ that we should use in Eq. (4) (of the current chapter)).

68

Feynman's Path Integral explained with basic Calculus

2. We evaluate Eq. (4) by the same technique: We replace k on the right of Eq. (15) of Chapter 3 by $-\frac{i}{\hbar}$ to obtain $K(t_b, x_b; t_a, x_a)$ in Eq. (4) (of the current chapter) as follows:

$$K(t_b, x_b; t_a, x_a) = \sqrt{\frac{m}{2\pi i \hbar (t_b - t_a)}} e^{\frac{im}{2\hbar} \frac{(x_b - x_a)^2}{(t_b - t_a)}} \quad \dots \dots \dots \dots \dots \text{Eq. (5)}.$$

[We used $-i = \frac{1}{i}$ in deriving Eq. (5).] Note that the argument of the exponential in Eq. (5) has the free-particle-action calculated for the least path or classical path, as given by Eq. (23) from Chapter 2. So, $K(t_b, x_b; t_a, x_a) \sim (constant) e^{\frac{i}{\hbar} S_0}$, implying that the action S_0 for the classical path continues to play an important role in the quantum world. We already explained in the context of Figure 22 how the classical paths and those close to it contribute most to the Path Integral. In Eq. (5), you can clearly see that from the presence of S_0 in the expression of the propagator. More about that later.

RECAP: **Feynman's Path Integrals**

The propagator $K(t_b, x_b; t_a, x_a)$ is defined as Feynman's Path Integral as follows:

$$K(t_b, x_b; t_a, x_a) = \lim_{N \to \infty} \left(\frac{1}{A}\right)^N \int_{-\infty}^{+\infty} dx_1 \int_{-\infty}^{+\infty} dx_2 \dots \int_{-\infty}^{+\infty} dx_{N-1} \, e^{\frac{i}{\hbar} S},$$

where S is the action ("discretized"), and $\frac{1}{A} = \sqrt{\frac{m}{2\pi i \hbar \epsilon}}$ ($\epsilon = \frac{t_b - t_a}{N}$), which applies for the free particle action, and other actions as well.

The propagator in Eq. (5) correctly gives the probability amplitude of a free particle to go from x_a at time t_a to x_b at time t_b. You could have evaluated Eq. (4) with appropriate $\frac{1}{A}$, by directly carrying out Gaussian integrations like

Chapter 4: Path Integrals in Quantum Mechanics

the ones that led you to Eq. (15) of Chapter 3. The calculations would have been exactly similar to those of Chapter 3 except that you would have dealt with complex exponential functions instead of real ones, and hence would have used the integral given by Eq. (17) of Chapter 1, instead of using Eq. (16) of Chapter 1, as you do in Chapter 3.

Incidentally, the Path Integral of Eq. (2) (of the current chapter) is sometimes written in the following equivalent form:

$K(t_b, x_b; t_a, x_a)$

$$= \lim_{N \to \infty} \frac{1}{A} \int_{-\infty}^{+\infty} \frac{dx_{N-1}}{A} \int_{-\infty}^{+\infty} \frac{dx_{N-2}}{A} \ldots \ldots \int_{-\infty}^{+\infty} \frac{dx_2}{A} \int_{-\infty}^{+\infty} \frac{dx_1}{A} e^{\frac{i}{\hbar}S} \ldots \ldots \text{ Eq. (6).}$$

Note that in Eq. (6), each dx multiplies an $\frac{1}{A}$ and that a lone $\frac{1}{A}$ stands in front of the multiple integral (You can count that there is a total N of $\frac{1}{A}$'s).

Next, we will try to understand Eq. (4) better. You might say: Wait a minute! We already evaluated Eq. (4), the answer of which is given by Eq. (5). What more is there to understand?

We haven't yet specifically discussed propagators defined over small time-intervals, a concept which will be crucial in our subsequent discussions. To remedy that, we take a closer look at Eq. (4) in the following.

Let us first revisit how we obtained S in Eq. (3). Pictorially speaking, we connected the space-time points (t_a, x_a), $(t_1, x_1), (t_2, x_2), \ldots, (t_b, x_b)$ with small line segments. (See Figure 23, which is the same as Figure 13 of the previous chapter.) Then, we calculated

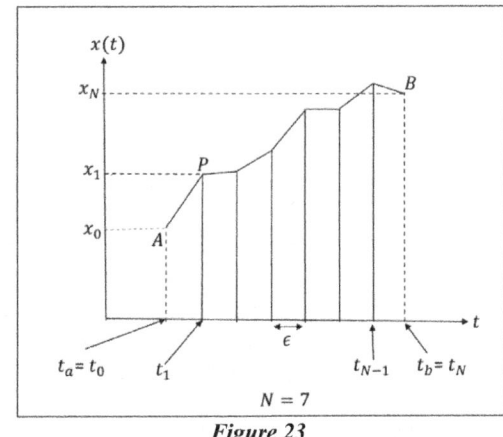

Figure 23

Feynman's Path Integral explained with basic Calculus

the slopes of the individual line-segments to discretize $\frac{dx(t)}{dt}$ for the respective time-intervals. For example, the slope of the i_{th} line-segment having the end-points (t_i, x_i) and (t_{i+1}, x_{i+1}) is $\frac{x_{i+1}-x_i}{\epsilon}$, ϵ being the length of the time-interval: $t_{i+1} - t_i = \epsilon$. (i is any number between 0 and $N-1$) (See Figure 24). In that time interval, the particle's speed is the line-segment's slope $\frac{x_{i+1}-x_i}{\epsilon}$, hence the particle's kinetic energy is:

$$KE = \frac{1}{2}m(speed)^2$$
$$= \frac{1}{2}m\left(\frac{x_{i+1}-x_i}{\epsilon}\right)^2$$
$$= \frac{m}{2}\frac{(x_{i+1}-x_i)^2}{\epsilon^2}.$$

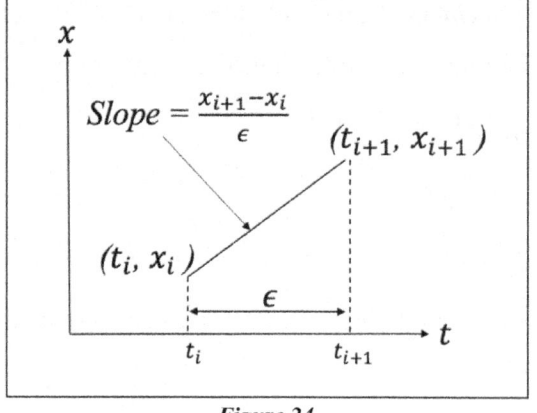

Figure 24

The particle's potential energy $PE = 0$ in the interval (t_i, t_{i+1}), since we are dealing with a free particle. Hence, the action S_i for the i_{th} line segment is:

$$S_i = \int_{t_i}^{t_{i+1}}(KE - PE)dt \text{ [Note that the lower and the upper limits of the integral are the beginning and end points of the } i_{th} \text{ time-interval.]}$$

$$= \int_{t_i}^{t_{i+1}} KE dt \text{ (Since } PE = 0\text{)}$$

$$\approx KE.\epsilon \text{ [Since } t_{i+1} - t_i = \epsilon \text{]}$$

$$= \frac{m}{2}\frac{(x_{i+1}-x_i)^2}{\epsilon^2}.\epsilon \text{ [Since } KE = \frac{m}{2}\frac{(x_{i+1}-x_i)^2}{\epsilon^2}\text{]}$$

$$= \frac{m}{2}\frac{(x_{i+1}-x_i)^2}{\epsilon} \quad \ldots \ldots \text{ Eq. (7)}.$$

Adding up such S_i's as given by Eq. (7) for all the line-segments (i varying from 0 to N-1) produces the discretized free-particle action S given by Eq. (3),

Chapter 4: Path Integrals in Quantum Mechanics

which we use in Eq. (2) and then apply the rule $e^{p_1+p_2+\cdots} = e^{p_1}e^{p_2}\ldots$ As a result, Eq. (2) assumes the following form:

$$K(t_b, x_b; t_a, x_a)$$
$$= \lim_{N\to\infty} \int_{-\infty}^{+\infty} dx_{N-1}\ldots dx_1 \left[\frac{1}{A}e^{\frac{i}{\hbar}\left(\frac{m(x_b-x_{N-1})^2}{2\epsilon}\right)}\right]\ldots\left[\frac{1}{A}e^{\frac{i}{\hbar}\left(\frac{m(x_{i+1}-x_i)^2}{2\epsilon}\right)}\right]$$
$$\ldots\left[\frac{1}{A}e^{\frac{i}{\hbar}\left(\frac{m(x_1-x_a)^2}{2\epsilon}\right)}\right]$$
.... Eq. (8).

In Eq. (8), the contribution of an i_{th} line segment to the Path Integral is:

$\frac{1}{A}e^{\frac{i}{\hbar}\left(\frac{m(x_{i+1}-x_i)^2}{2\epsilon}\right)} = \frac{1}{A}e^{\frac{i}{\hbar}S_i}$, where S_i is given by Eq. (7). $\frac{1}{A}e^{\frac{i}{\hbar}\left(\frac{m(x_{i+1}-x_i)^2}{2\epsilon}\right)}$ is precisely the infinitesimal-time free particle propagator $K(t_{i+1}, x_{i+1}; t_i, x_i)$, i.e., the probability amplitude for the free particle to go from (t_i, x_i) to (t_{i+1}, x_{i+1}).

$$K(t_{i+1}, x_{i+1}; t_i, x_i) = \frac{1}{A}e^{\frac{i}{\hbar}\left(\frac{m(x_{i+1}-x_i)^2}{2\epsilon}\right)}$$ .... Eq. (9).

Note that in the Path Integral of Eq. (8), there are N infinitesimal-time propagators corresponding to N line-segments like the one shown in Figure 24. Also, note that since the total number of line-segments is N, you get a net multiplicative factor of $\left(\frac{1}{A}\right)^N$, the standard normalization factor for a Path Integral.

Finite-time Versus Infinitesimal-time Free Particle Propagators

Let us compare the finite-time free particle propagator with an infinitesimal-time one. The finite-time propagator between the space-time points $A(t_a, x_a)$ and $B(t_b, x_b)$ (as given by Eq. (5)) is the particle's probability amplitude to go from A to B. The jagged path in Figure 23 (reproduced in this section for

Feynman's Path Integral explained with basic Calculus

convenience), which connects A and B, consists of the points $(t_a, x_a), (t_1, x_1), \ldots, (t_{N-1}, x_{N-1}), (t_b, x_b)$. As we discussed before, you get different paths by changing the values of one or more intermediate position variables $x_1, x_2, \ldots x_{N-1}$, corresponding to the intermediate time-points $t_1, t_2, \ldots t_{N-1}$ in the time-interval (t_a, t_b). All such paths contribute to the calculation of the free particle's probability amplitude to go from A to B, i.e., the propagator $K(t_b, x_b; t_a, x_a)$ (as given by Eq. (4)).

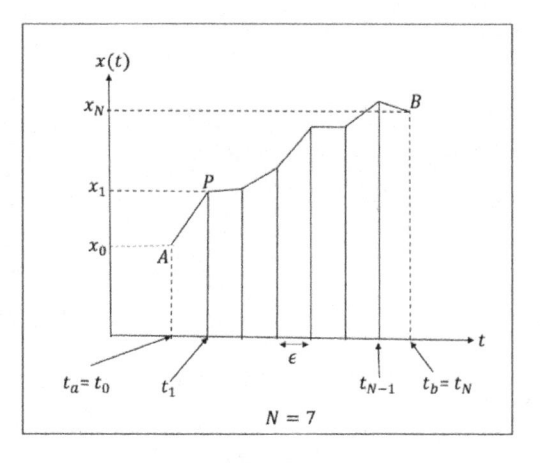

But, for an infinitesimal time-interval, viz., (t_i, t_{i+1}), as shown in Figure 24 (reproduced here for convenience), you only have "one path" (a straight line-segment) defined in that interval, since you have no "intermediate" time-point and hence no intermediate x variable. Hence, the infinitesimal-time propagator (defined for the infinitesimal interval (t_i, t_{i+1}) and given by Eq. (9), is determined by the single line-segment of Figure 24 (reproduced). (Of course, you can divide the interval (t_i, t_{i+1}) into more time-intervals and thus "create" intermediate x variables, but then you will face a similar situation w.r.t the smallest interval in your new scheme.)

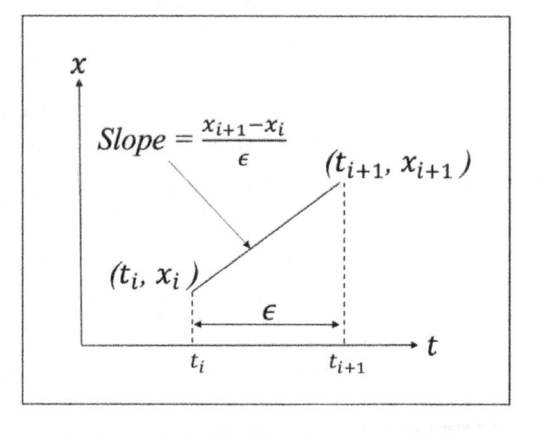

Chapter 4: Path Integrals in Quantum Mechanics

This completes our theoretical discussion on free particle propagators. In the next section, we will show how a free particle propagator can be used to derive De Broglie's wavelength and Einstein's energy frequency relationship. Free particle propagators are tremendously helpful even when particles aren't free, i.e., they move in force fields and have potential energies, which we will briefly touch upon in Appendix 4.

We end this chapter by demonstrating how to obtain the probability density function from the probability amplitude. Assuming that the free particle starts its journey at $x = x_a$ at time $t = t_a$, we are interested in the probability density of finding the particle at time $t = t_b$ at the location $x = x_b$. Per the rule of quantum mechanics, taking the absolute square of the free particle propagator

$$K(t_b, x_b; t_a, x_a) = \sqrt{\frac{m}{2\pi i \hbar (t_b - t_a)}} e^{\frac{im}{2\hbar} \frac{(x_b - x_a)^2}{(t_b - t_a)}}$$ (Eq. (5)), we get the required probability density $P(t_b, x_b; t_a, x_a) = \frac{m}{2\pi \hbar (t_b - t_a)}$. As you can see, $P(t_b, x_b; t_a, x_a)$ does not depend on the space-variable, which implies that although the particle starts at x_a at time t_a, it is equally likely to find the particle at any spatial location at a later time t_b.

Derivation of De Broglie's and Einstein's Relation from Free Particle Propagator

In this section, we will derive de Broglie's wavelength from the free particle propagator $K(t_b, x_b; t_a, x_a) = \sqrt{\frac{m}{2\pi i \hbar (t_b - t_a)}} e^{\frac{im}{2\hbar} \frac{(x_b - x_a)^2}{(t_b - t_a)}}$ (Given by Eq. (5)). Without any loss of generality, we will set $x_a = 0$, $t_a = 0$ and $x_b = x$, $t_b = t$ in the above expression of the propagator to obtain $K(t, x; 0, 0) = $

Feynman's Path Integral explained with basic Calculus

$\sqrt{\dfrac{m}{2\pi i \hbar t}}\, e^{\dfrac{imx^2}{2\hbar t}}$, which gives the probability amplitude of the particle going from $(0,0)$ to (t,x) in the space-time diagram.

We plot the real part of $\sqrt{iK}(t,x;0,0)$ as a function of x for a fixed t in Figure 25. The graph represents a periodic function in the space variable x. The distance between, say, the nodes, of the periodic function stays relatively constant over a few cycles when x is large. So, to calculate the wavelength or the spatial period λ, we assume x to be large and replace x in $K(t,x;0,0)$ by $x+\lambda$ to get $K(t,x+\lambda;0,0)$, which we will set equal to $K(t,x;0,0)$: $K(t,x;0,0) = K(t,x+\lambda;0,0)$. Thus, we obtain:

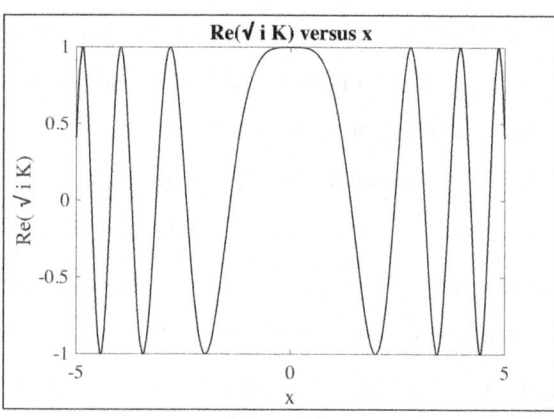

Figure 25

$$\sqrt{\dfrac{m}{2\pi i \hbar t}}\, e^{\dfrac{imx^2}{2\hbar t}} = \sqrt{\dfrac{m}{2\pi i \hbar t}}\, e^{\dfrac{im(x+\lambda)^2}{2\hbar t}}.$$

Cancelling out the common factor $\sqrt{\dfrac{m}{2\pi i \hbar t}}$ from both sides of the above equation and then multiplying both sides by $e^{-\dfrac{imx^2}{2\hbar t}}$, we obtain:

$$e^{i\left[\dfrac{m(x+\lambda)^2}{2\hbar t} - \dfrac{mx^2}{2\hbar t}\right]} = 1.$$

Since, $e^{i(2\pi)} = 1$, we can write from the above:

$$\dfrac{m(x+\lambda)^2}{2\hbar t} - \dfrac{mx^2}{2\hbar t} = 2\pi.$$

Now, expanding the square on the left,

$$\dfrac{m(x^2+2x\lambda+\lambda^2)}{2\hbar t} - \dfrac{mx^2}{2\hbar t} = 2\pi.$$

Chapter 4: Path Integrals in Quantum Mechanics

Per our assumption, x is much larger than λ and hence ignoring λ^2 in the first term on left side of the above equation, we get after some further algebra:

$$m\frac{x}{t}\frac{\lambda}{\hbar} = 2\pi \quad \text{... Eq. (10).}$$

But, from the classical perspective, $\frac{x}{t}$ is the particle's speed v [The classical perspective is made possible because of x (the dimension of the physical system) being much larger than the wavelength λ]. So, on the left of Eq. (10), you have mv, the particle's momentum, times $\frac{\lambda}{\hbar}$: $mv\frac{\lambda}{\hbar}$. Writing p for mv, we obtain from Eq. (10):

$$\lambda = \frac{h}{p}, \text{ [We used } \hbar = \frac{h}{2\pi}].$$

The above is the famous de Broglie relation, connecting the wavelength λ of the quantum wave associated with a particle of mass m to the particle's momentum p.

Next, we plot the real part of $\sqrt{iK}(t,x;0,0)$ as a function of time, keeping x fixed (Figure 26). As you can see from the graph, the amplitude of the plotted function falls off with increasing t, but when t is large, the

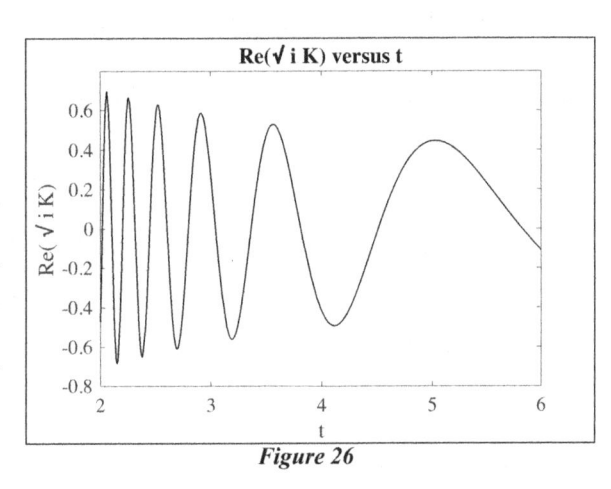

Figure 26

amplitudes are relatively constant over a few cycles and the plotted function behaves as a fixed amplitude sinusoid. So, we can say that $K(t,x;0,0)$ approximately repeats itself when t is large. Let the time period of K is T (which we assume to be much less than t.) Hence,

Feynman's Path Integral explained with basic Calculus

$$\sqrt{\frac{m}{2\pi i \hbar t}} e^{\frac{imx^2}{2\hbar t}} = \sqrt{\frac{m}{2\pi i \hbar (t+T)}} e^{\frac{imx^2}{2\hbar(t+T)}} \quad \ldots\ldots\ldots\ldots \text{Eq. (11).}$$

Since $t \gg T$, the pre-factors $\sqrt{\frac{m}{2\pi i \hbar t}}$ and $\sqrt{\frac{m}{2\pi i \hbar (t+T)}}$ are approximately equal and can be cancelled out.

Then, using $e^{i(2\pi)} = 1$, as we did to derive Eq. (10), we obtain the following from Eq. (11):

$$\frac{mx^2}{2\hbar t} - \frac{mx^2}{2\hbar(t+T)} = 2\pi.$$

In the above equation, writing $\frac{1}{t+T} = \frac{1}{t\left(1+\frac{T}{t}\right)} = \frac{1}{t}\left(1+\frac{T}{t}\right)^{-1} \approx \frac{1}{t}\left(1-\frac{T}{t}\right)$

(Taylor expansion), and after some further algebra, we obtain:

$$\frac{mx^2}{2\hbar t^2} T = 2\pi \quad \ldots\ldots\ldots\ldots\ldots\ldots\ldots\ldots\ldots\ldots\ldots\ldots\ldots\ldots \text{Eq. (12).}$$

But, since $\frac{x}{t}$ is the particle's speed v (from the classical perspective), $\frac{1}{2}\frac{mx^2}{t^2}$ in Eq. (12) is $\frac{1}{2}mv^2$, which is the classical kinetic energy of the particle. Since the particle is free, its kinetic energy is also its total energy E. Hence, $\frac{1}{2}\frac{mx^2}{t^2} = E$, using which in Eq. (12), we get $E = \frac{h}{T} = hf$ (where f is the frequency given by $f = \frac{1}{T}$.) $E = hf$ is the famous relation from Einstein, connecting the energy of a particle to the frequency of the quantum wave associated with a massive particle.

Chapter 5: Path Integrals in the Presence of Potential Energies

In this chapter, we will discuss Path Integrals when the particle is not free but has a potential energy $PE = V(x)$. This means that the action S, used in $e^{i\frac{S}{\hbar}}$ in the Path Integral, will have a potential energy term in addition to the kinetic energy (KE) term. (In the case of the free particle, we only had a KE term in the action, the particle's PE being zero.) But before getting into the details of such Path Integrals, let us briefly discuss the potential energies arising in the microscopic world.

Potential Energies in the Microscopic World

An electron of charge q ($q < 0$), moving in a constant electric field E along the direction of the field, feels a force qE (Figure 27). If the electron covers a distance y along the field's direction, the electron's electric potential energy is $-qEy$, which is linear in the distance variable, just like the gravitational potential energy of a falling stone (as discussed in Chapter 2). We will simply denote the electron's potential energy, linear in y, as $PE = -ky$, k being a constant. Hence, the action for an electron in a constant electric field is given by: $S =$

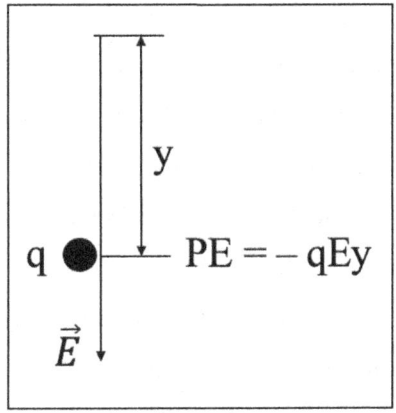

Figure 27

Feynman's Path Integral explained with basic Calculus

$\int_{t=t_a}^{t=t_b}[KE - PE]dt = \int_{t=t_a}^{t=t_b}\left[\frac{m}{2}\left(\frac{dy(t)}{dt}\right)^2 + ky\right]dt$, which is similar to Eq. (24) of Chapter 2.

Next, another potential energy, frequently encountered in the microscopic world, is formally similar to the potential energy of a spring-mass system ($PE = \frac{1}{2}kx^2$) [See Figure 12 in Chapter 2]. For example, consider atoms being arranged periodically at the lattice sites of a crystal. Atoms behave as if connected by little springs and hence their vibrations involve a potential energy that is a quadratic function of the position variables of the atoms. The collective oscillation of the atoms, also called phonons, are responsible for carrying sound through a material. An object of mass m with a potential energy quadratic in the position (space) variable embodies one of the very basic quantum mechanical problems; the corresponding action being given by $S = \int_{t=t_a}^{t=t_b}[KE - PE]dt = \int_{t=t_a}^{t=t_b}\left[\frac{m}{2}\left(\frac{dx(t)}{dt}\right)^2 - \frac{1}{2}kx^2\right]dt$.

The quadratic potential energy $\frac{1}{2}kx^2$ is also called the harmonic potential energy, since, classically, an object with the quadratic potential energy executes harmonic (oscillatory) motion, which you proved in Exercise 4, Chapter 2. Harmonic potential energy also arises in the quantum mechanical treatment of electromagnetic radiations, the discussion of which is beyond the scope of the book.

Yet another potential energy, that is of importance in quantum mechanics, is the Coulomb potential energy, present between two charges separated by a distance. For example, in a hydrogen atom, an electron (of charge q), located at a distance r from a positively charged nucleus (of charge Q), has the potential energy $PE = k\frac{qQ}{r}$, where k is a constant. We will not be using this potential energy in this book. We will stick only to the linear and the quadratic potential energies, viz., $PE = -kx$ and $PE = \frac{1}{2}kx^2$.

Chapter 5: Path Integrals in the Presence of Potential Energies

Path Integrals with the Potential Energy Term in the Action

Next, we will discuss how to calculate the Path Integral, when the action has a potential energy term.

The action for a particle of mass m with an arbitrary, time-independent, potential energy $V(x)$ is given by: $S = \int_{t_a}^{t_b}(KE - PE)dt = \int_{t=t_a}^{t=t_b}\left[\frac{m}{2}\left(\frac{dx(t)}{dt}\right)^2 - V(x)\right]dt$. And the probability amplitude for that same particle to go from the point $A(t_a, x_a)$ to the point $B(t_b, x_b)$ in the x-t plane is given by:

$K(t_b, x_b; t_a, x_a)$
$= \lim_{N \to \infty} \left(\frac{1}{A}\right)^N \int_{-\infty}^{+\infty} dx_{N-1} \int_{-\infty}^{+\infty} dx_{N-2} \ldots \ldots \int_{-\infty}^{+\infty} dx_2 \int_{-\infty}^{+\infty} dx_1 \, e^{\frac{i}{\hbar}S}$

.... Eq. (1),

where, $S = \int_{t=t_a}^{t=t_b}\left[\frac{m}{2}\left(\frac{dx(t)}{dt}\right)^2 - V(x)\right]dt$.

Like before, the time interval (t_a, t_b) is divided into N equal parts by the time-points $t_1, t_2, t_3, \ldots\ldots t_{N-1}$, the corresponding x-values in the $x - t$ plane being $x(t_1) = x_1, x(t_2) = x_2, \ldots, x(t_{N-1}) = x_{N-1}$. The endpoints of the time-interval are: $t_0 = t_a$ and $t_N = t_b$, and the x-values at t_a and t_b are $x(t_a) = x_a$ and $x(t_b) = x_b$ respectively.

Feynman's Path Integral explained with basic Calculus

To calculate Eq. (1), you need to discretize S, as we did in case of a free particle. But before we get to that, what is $\frac{1}{A}$ in Eq. (1)? As mentioned before, $\frac{1}{A}$ in Eq. (1) is the same as in the free particle case, a result we will prove later in Chapter 6. How does the discretization play out when a general potential energy $V(x)$ is present? To explain that, once again, in Figure 28, we consider a small segment in the ith time interval (t_i, t_{i+1}) of the jagged path as in Figure 23 and calculate the action for that tiny piece of path. (i can represent any of the N time intervals that (t_a, t_b) is divided into.)

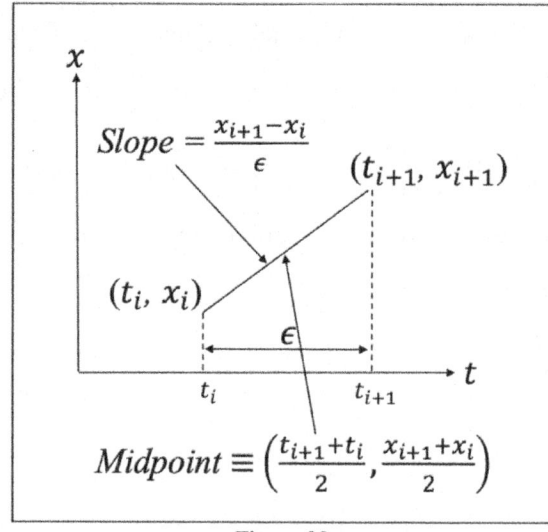

Figure 28

As shown in Figure 28, the endpoints of the ith segment are (t_i, x_i) and (t_{i+1}, x_{i+1}). Calculating the particle's kinetic energy for that segment is exactly similar to the free-particle case. The speed ($v = \frac{dx}{dt}$) is simply the slope of the line-segment that joins the points, i.e., $v = \frac{x_{i+1}-x_i}{\epsilon}$, and hence the $KE = \frac{1}{2}mv^2 = \frac{1}{2}m\left(\frac{x_{i+1}-x_i}{\epsilon}\right)^2 = \frac{m}{2}\frac{(x_{i+1}-x_i)^2}{\epsilon^2}$. The particle's potential energy for the same line-segment can be evaluated at the segment's mid-point. Why the midpoint? Well, why not? It looks reasonable. Since the co-ordinates of the midpoint of the line-segment are $\left(\frac{t_i+t_{i+1}}{2}, \frac{x_i+x_{i+1}}{2}\right)$, we will assign the time-independent potential energy $V = V\left(\frac{x_{i+1}+x_i}{2}\right)$ for the line-segment in the

Chapter 5: Path Integrals in the Presence of Potential Energies

interval (t_i, t_{i+1}). (Note that V, being time-independent, depends only on the space variables.) Hence the action for the i_{th} line-segment is:

$$S_i = \int_{t_i}^{t_{i+1}}(KE - PE)dt$$
$$\approx (KE - PE)\epsilon \text{ [Since } t_{i+1} - t_i = \epsilon]$$
$$= \frac{m}{2}\frac{(x_{i+1}-x_i)^2}{\epsilon} - \epsilon V\left(\frac{x_{i+1}+x_i}{2}\right) \text{ [Using } KE = \frac{m}{2}\frac{(x_{i+1}-x_i)^2}{\epsilon^2} \text{ and}$$
$$PE = V\left(\frac{x_{i+1}+x_i}{2}\right)].$$

Adding up such contributions S_i's from all the line-segments ($i = 0, 1, \ldots, N-1$), we get the complete discretized action as follows:

$$S = \left[\frac{m}{2}\frac{(x_1-x_a)^2}{\epsilon} - \epsilon V\left(\frac{x_1+x_a}{2}\right)\right] + \left[\frac{m}{2}\frac{(x_2-x_1)^2}{\epsilon} - \epsilon V\left(\frac{x_1+x_2}{2}\right)\right] + \cdots$$
$$\cdots + \left[\frac{m}{2}\frac{(x_b-x_{N-1})^2}{\epsilon} - \epsilon V\left(\frac{x_b+x_{N-1}}{2}\right)\right] \quad\ldots\ldots\ldots\ldots\ldots \text{Eq. (2)}.$$

After using S (as given by Eq. (2)) in Eq. (1) and then applying the rule $e^{p_1+p_2+\cdots} = e^{p_1}e^{p_2}\ldots$, we obtain from Eq. (1),

$$K(t_b, x_b; t_a, x_a) = \lim_{N\to\infty}$$

$$\int_{-\infty}^{+\infty} dx_{N-1}\ldots dx_2 dx_1 \left[\frac{1}{A}e^{\frac{i}{\hbar}\left(\frac{m(x_1-x_a)^2}{2\epsilon} - \epsilon V\left(\frac{x_a+x_1}{2}\right)\right)}\right]\ldots\left[\frac{1}{A}e^{\frac{i}{\hbar}\left(\frac{m(x_{i+1}-x_i)^2}{2\epsilon} - \epsilon V\left(\frac{x_{i+1}+x_i}{2}\right)\right)}\right]$$

$$\ldots\ldots\ldots product\ of\ N\ terms \qquad \ldots\ldots\ldots\ldots\text{Eq. (3)}.$$

Note that Eq. (3) is similar to Eq. (8) of Chapter 4, except that the potential energy V is present in Eq. (3) of the current Chapter but absent from Eq. (8) of Chapter 4.

The contribution of a generic i_{th} line segment (Figure 28) to the Path Integral in Eq. (3) is: $\frac{1}{A}e^{\frac{i}{\hbar}(KE-PE)\epsilon} = \frac{1}{A}e^{\frac{i}{\hbar}\left(\frac{m(x_{i+1}-x_i)^2}{2\epsilon} - \epsilon V\left(\frac{x_{i+1}+x_i}{2}\right)\right)}$, which is precisely the propagator $K(t_{i+1}, x_{i+1}; t_i, x_i)$, i.e., the probability amplitude for the

Feynman's Path Integral explained with basic Calculus

particle to go from (t_i, x_i) to (t_{i+1}, x_{i+1}) in the presence of a force-field, i.e., when the particle has both kinetic and potential energy. Hence:

$$K(t_{i+1}, x_{i+1}; t_i, x_i) = \frac{1}{A} e^{\frac{i}{\hbar}\left(\frac{m(x_{i+1}-x_i)^2}{2\epsilon} - \epsilon V\left(\frac{x_{i+1}+x_i}{2}\right)\right)} \quad \ldots\ldots\ldots\ \text{Eq. (4)}.$$

There are N of these infinitesimal-time propagators for N line-segments. [Note that since there are N such line segments, you get a net multiplicative factor of $\left(\frac{1}{A}\right)^N$, the standard normalization factor for a Path Integral.]

$\frac{1}{A}$, in Eq. (4), is the same as in the case for the free particle, which we will prove later while deriving Schrodinger's equation using the infinitesimal-time propagator in Chapter 6.

As you can probably guess, evaluating Eq. (3) for a general potential energy can be enormously complicated. Even for a potential energy varying quadratically with the position or the space variable, a direct calculation of Eq. (3) is pretty involved, which we will not get into. Instead, we will use the indirect method for calculating the Path Integral, when the potential energy is present. Before we get to that, we will first prove a very important property of propagators in the next section, which we will need to derive Schrodinger's equation by using the propagator in Chapter 6.

Chapter 5: Path Integrals in the Presence of Potential Energies

---

RECAP: <u>Infinitesimal-time Propagator with Potential Energy</u>

Infinitesimal-time Propagator:

$$K(t_{i+1}, x_{i+1}; t_i, x_i) = \frac{1}{A} e^{\frac{i}{\hbar}\left(\frac{m(x_{i+1}-x_i)^2}{2 \epsilon} - \epsilon V\left(\frac{x_{i+1}+x_i}{2}\right)\right)}$$

Slope $= \dfrac{x_{i+1}-x_i}{\epsilon}$

(t_{i+1}, x_{i+1})

(t_i, x_i)

$\text{Midpoint} \equiv \left(\dfrac{t_{i+1}+t_i}{2}, \dfrac{x_{i+1}+x_i}{2}\right)$

Multiplicative Law of Propagators

$A(t_a, x_a)$, $B(t_b, x_b)$ and $C(t_c, x_c)$ are three points as shown in Figure 29. t_c is in between t_a and t_b.

The propagator $K(t_b, x_b; t_a, x_a)$ is the probability amplitude for the particle to go from A to B.

$K(t_b, x_b; t_c, x_c)$ is the probability amplitude for the particle to go from C to B.

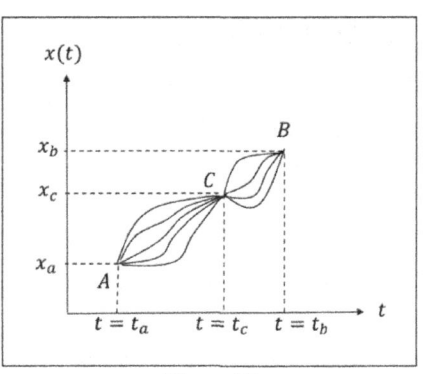

Figure 29

$K(t_c, x_c; t_a, x_a)$ is the probability amplitude for the particle to go from A to C.

84

Feynman's Path Integral explained with basic Calculus

In this section, we will prove that the above propagators satisfy:

$$K(t_b, x_b; t_a, x_a) = \int_{-\infty}^{+\infty} dx_c K(t_b, x_b; t_c, x_c) K(t_c, x_c; t_a, x_a) \dots \dots \dots \text{Eq. (5)}.$$

We refer to Eq. (5) as the "multiplicative law of propagators". Denoting (t_b, x_b) as B, (t_c, x_c) as C, and (t_a, x_a) as A, we can express $K(t_b, x_b; t_c, x_c) \equiv K(B, C)$, $K(t_c, x_c; t_a, x_a) \equiv K(C, A)$ and $K(t_b, x_b; t_a, x_a) \equiv K(B, A)$. Using these notations, Eq. (5) can be written as:

$$K(B, A) = \int_{-\infty}^{+\infty} dx_c K(B, C) K(C, A) \dots \dots \dots \dots \text{Eq. (6)}.$$

Eq. (6) will come in handy in the next chapter while deriving Schrodinger's equation. In Figure 29, multiple paths are drawn to represent $K(C, A)$, which, by definition, involves summing over the paths between the end-points C and A. Similarly, multiple paths between B and C are used to represent $K(B, C)$. So, in essence, you split the sum over all paths from A to B into the sum over all paths from A to an intermediate point C, and the sum over all paths from C to the final point B. Finally, you consider all possible intermediate points C (i.e., integrate w.r.t x_c) as shown in Eq. (6).

Next we derive Eq. (5).

Proof of Eq. (5)

As usual, the time interval (t_a, t_b) is divided into N equal parts by the time-points: $t_1, t_2, \dots, t_{N-1}$. Let, t_c be the j_{th} time point t_j as shown in Figure 30. Since, $t_c \equiv t_j$, the x-variable corresponding to t_c is $x_c \equiv x_j$. Let, the point

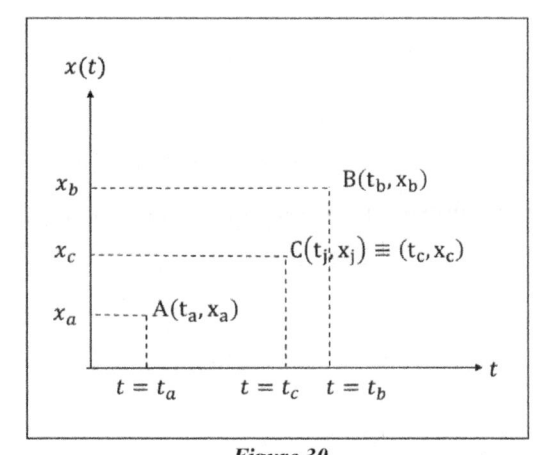

Figure 30

Chapter 5: Path Integrals in the Presence of Potential Energies

$(t_c, x_c) \equiv (t_j, x_j)$ be denoted by C.

We consider the action S with a potential energy (as given by Eq. (2)). We can split S into two parts as shown in the following: the first part beginning with x_a and ending with x_j, and the second part beginning with x_j and ending with x_b as follows (follow the outermost first brackets):

$$S = \left(\left[\frac{m}{2}\frac{(x_1-x_a)^2}{\epsilon} - \epsilon V\left(\frac{x_1+x_a}{2}\right)\right] + \cdots + \left[\frac{m}{2}\frac{(x_j-x_{j-1})^2}{\epsilon} - \epsilon V\left(\frac{x_{j-1}+x_j}{2}\right)\right]\right)$$

$$+ \left(\left[\frac{m}{2}\frac{(x_{j+1}-x_j)^2}{\epsilon} - \epsilon V\left(\frac{x_j+x_{j+1}}{2}\right)\right] + \cdots\right.$$

$$\left. + \left[\frac{m}{2}\frac{(x_b-x_{N-1})^2}{\epsilon} - \epsilon V\left(\frac{x_b+x_{N-1}}{2}\right)\right]\right)$$

Let us denote the first part of the above action as $S(x_j; x_a)$: the final point x_j written before the initial point x_a. Similarly, the second part of S be denoted by $S(x_b; x_j)$. Hence,

$S = S(x_j; x_a) + S(x_b; x_j)$, where:

$$S(x_j; x_a) = \left[\frac{m}{2}\frac{(x_1-x_a)^2}{\epsilon} - \epsilon V\left(\frac{x_1+x_a}{2}\right)\right] + \cdots + \left[\frac{m}{2}\frac{(x_j-x_{j-1})^2}{\epsilon} - \epsilon V\left(\frac{x_{j-1}+x_j}{2}\right)\right]$$

$$\dots\dots\dots\dots\dots\dots\dots\dots\dots\dots \text{Eq. (7),}$$

and

$$S(x_b; x_j)$$

$$= \left[\frac{m}{2}\frac{(x_{j+1}-x_j)^2}{\epsilon} - \epsilon V\left(\frac{x_j+x_{j+1}}{2}\right)\right] + \cdots + \left[\frac{m}{2}\frac{(x_b-x_{N-1})^2}{\epsilon} - \epsilon V\left(\frac{x_b+x_{N-1}}{2}\right)\right]$$

$$\dots\dots\dots\dots\dots\dots\dots\dots\dots \text{Eq. (8).}$$

Using $S = S(x_j; x_a) + S(x_b; x_j)$, as given by the above equations, in $e^{\frac{i}{\hbar}S}$, we get, $e^{\frac{i}{\hbar}S} = e^{\frac{i}{\hbar}(S(x_j;x_a)+S(x_b;x_j))} = e^{\frac{i}{\hbar}S(x_j;x_a)} e^{\frac{i}{\hbar}S(x_b;x_j)}$, which we use in the Path Integral in Eq. (1) to obtain:

Feynman's Path Integral explained with basic Calculus

$$K(t_b, x_b; t_a, x_a)$$
$$= \lim_{N \to \infty} \left(\frac{1}{A}\right)^N \int_{-\infty}^{+\infty} dx_{N-1} \ldots$$
$$\ldots \int_{-\infty}^{+\infty} dx_{j+1} \int_{-\infty}^{+\infty} dx_j \int_{-\infty}^{+\infty} dx_{j-1} \ldots \int_{-\infty}^{+\infty} dx_1 \, e^{\frac{i}{\hbar}S(x_b; x_j)} e^{\frac{i}{\hbar}S(x_j; x_a)}$$

We rewrite the above expression in a slightly different way, as shown in the following, the reason for which will be clear later.

$$K(t_b, x_b; t_a, x_a)$$
$$= \lim_{N \to \infty} \frac{1}{A} \int_{-\infty}^{+\infty} \frac{dx_{N-1}}{A} \ldots$$
$$\ldots \int_{-\infty}^{+\infty} \frac{dx_{j+1}}{A} \int_{-\infty}^{+\infty} \frac{dx_j}{A} \int_{-\infty}^{+\infty} \frac{dx_{j-1}}{A} \ldots \int_{-\infty}^{+\infty} \frac{dx_2}{A} \int_{-\infty}^{+\infty} \frac{dx_1}{A} \, e^{\frac{i}{\hbar}S(x_b; x_j)} e^{\frac{i}{\hbar}S(x_j; x_a)}$$
$$\ldots\ldots\ldots\ldots\ldots\ldots\ldots\ldots\ldots \text{ Eq. (9).}$$

Note that in Eq. (9), every dx variable has an $\frac{1}{A}$ multiplying it. And there is a lone $\frac{1}{A}$ sitting in front of the multiple integral. By counting the number $\frac{1}{A}$'s, you can convince yourself that there is a total of N of them, as expected. Next we rewrite Eq. (9) in the following manner:

$$K(t_b, x_b; t_a, x_a)$$
$$= \lim_{N \to \infty} \left(\frac{1}{A} \int_{-\infty}^{+\infty} \frac{dx_{N-1}}{A} \ldots \int_{-\infty}^{+\infty} \frac{dx_{j+1}}{A} e^{\frac{i}{\hbar}S(x_b; x_j)}\right) \int_{-\infty}^{+\infty} dx_j$$
$$\left(\frac{1}{A} \int_{-\infty}^{+\infty} \frac{dx_{j-1}}{A} \ldots \int_{-\infty}^{+\infty} \frac{dx_2}{A} \int_{-\infty}^{+\infty} \frac{dx_1}{A} e^{\frac{i}{\hbar}S(x_j; x_a)}\right)$$
$$\ldots\ldots\ldots\ldots\ldots\ldots\ldots\ldots\ldots \text{ Eq. (10).}$$

Note how in Eq. (10), dx_j is the only integration variable without an $\frac{1}{A}$ factor with it. We kept the $\frac{1}{A}$'s with other dx's. Once again, you can convince yourself, by counting, that the total number of $\frac{1}{A}$'s in Eq. (10) is N. Here is a way to bypass the rigorous counting and still convince yourself that this is correct: In Eq. (9), every dx variable is multiplied by $\frac{1}{A}$, and there is a dangling $\frac{1}{A}$ in the front. Hence, if we only strip dx_j of an accompanying $\frac{1}{A}$, we will have

Chapter 5: Path Integrals in the Presence of Potential Energies

two dangling $\frac{1}{A}$'s (one already existed and we got one by freeing dx_j of $\frac{1}{A}$). They show up in Eq. (10), one each, as dangling $\frac{1}{A}$, in the bracketed expressions to the left and to the right of dx_j respectively. Now:

$$\frac{1}{A}\int_{-\infty}^{+\infty}\frac{dx_{N-1}}{A}\ldots\ldots\int_{-\infty}^{+\infty}\frac{dx_{j+1}}{A}e^{\frac{i}{\hbar}S(x_b;x_j)} = K(t_b, x_b; t_j, x_j),$$ [Following the definition of a propagator as in Eq. (9). The intermediate variables in this case are: $x_{j+1}, x_{j+2}, \ldots x_{N-1}$]

and

$$\frac{1}{A}\int_{-\infty}^{+\infty}\frac{dx_{j-1}}{A}\ldots\ldots\int_{-\infty}^{+\infty}\frac{dx_2}{A}\int_{-\infty}^{+\infty}\frac{dx_1}{A}e^{\frac{i}{\hbar}S(x_j;x_a)} = K(t_j, x_j; t_a, x_a)$$

[The intermediate variables are: $x_1, x_2, \ldots x_{j-1}$.]

Hence, Eq. (10) can be expressed as:

$$K(t_b, x_b; t_a, x_a) = \int_{-\infty}^{+\infty} dx_j K(t_b, x_b; t_j, x_j) K(t_j, x_j; t_a, x_a)\ldots\ldots\ldots\text{Eq. (11)}.$$

Since (t_j, x_j) are the co-ordinates of the point C, we will rewrite (t_j, x_j) as (t_c, x_c). Hence, $K(t_b, x_b; t_j, x_j) = K(t_b, x_b; t_c, x_c)$ and $K(t_j, x_j; t_a, x_a) = K(t_c, x_c; t_a, x_a)$. Therefore, Eq. (11) is rewritten as:

$$K(t_b, x_b; t_a, x_a) = \int_{-\infty}^{+\infty} dx_c K(t_b, x_b; t_c, x_c) K(t_c, x_c; t_a, x_a),$$ which is nothing but Eq. (5). Thus Eq. (5) is proved.

You can generalize Eq. (6) (which is same as Eq. (5) in an alternative notation) for any number of intermediate variables. For example, you can show that:

$$K(B, A) = \int_{-\infty}^{+\infty} dx_c dx_d K(B, D) K(D, C) K(C, A),$$

where the time-points corresponding to C and D are t_c and t_d, where $t_b > t_d > t_c > t_a$, and $x(t_c) = x_c, x(t_d) = x_d$. The endpoints are $x(t_a) = x_a$ and $x(t_b) = x_b$ as usual.

Feynman's Path Integral explained with basic Calculus

We will not prove the above equation here, but the proof is along the same lines as the derivation of Eq. (5).

We derived Eq. (6) for the general case when potential energy is present. Hence Eq. (6) is also valid for a free particle which is a special case where potential energy equals zero; you use free particle propagators for the K's in Eq. (6) when dealing with free particles.

Examples of Calculating Path Integrals with Potential Energy

We now resume our conversation on evaluating the Path Integral involving a potential energy. In this section, we will consider the potential energies, linear and quadratic in the position variable, and use the indirect method introduced before (in Chapter 3, section: "Indirect Method for Evaluating Path Integrals") for evaluating Path Integrals. (You may want to take a quick glance at that section before continuing.)

Example 1 (Potential Energy is Linear in the Position Variable) We use $PE = -ky$ for the particle's potential energy, which, you can see, is linear in the position variable y. The particle's kinetic energy is: $\frac{m}{2}\left(\frac{dy(t)}{dt}\right)^2$. Hence, the action $S = \int_{t=t_a}^{t=t_b}[KE - PE]dt$ is explicitly written out as:

$$S = \int_{t=t_a}^{t=t_b}\left[\frac{m}{2}\left(\frac{dy(t)}{dt}\right)^2 + ky(t)\right]dt. \quad\quad\quad\text{Eq. (12)}.$$

The action in Eq. (12) is exactly similar to Eq. (24) of Chapter 2, where we used $PE = -mgy$, whereas for the current problem, $PE = -ky$; k in the current problem plays the role of mg in Chapter 2. We will evaluate the particle's probability amplitude (the propagator) $K(t_b, y_b; t_a, y_a)$ to go from

Chapter 5: Path Integrals in the Presence of Potential Energies

A (t_a, y_a) to B(t_b, y_b), where $K(t_b, y_b; t_a, y_a)$ is given by the standard expression (Eq. (1)) for the Path Integral:

$K(t_b, y_b; t_a, y_a) =$

$$\lim_{N \to \infty} \left(\frac{1}{A}\right)^N \int_{-\infty}^{+\infty} dy_{N-1} \int_{-\infty}^{+\infty} dy_{N-2} \cdots \int_{-\infty}^{+\infty} dy_3 \int_{-\infty}^{+\infty} dy_2 \int_{-\infty}^{+\infty} dy_1\, e^{\frac{i}{\hbar}S}$$

.... Eq. (13).

S in Eq. (13) is given by Eq. (12). Some of the steps to calculate the propagator will be common with calculating the "least path" or the "classical path" for the corresponding problem (Example 2 from Chapter 2), which is true about the indirect method in general, as you saw in Chapter 3, the only other time when you applied the indirect method to evaluate a Path Integral.

Let $\bar{y}(t)$ be the least path (for which S is a minimum) defined between two given points A and B as shown in the figure on the right side. An arbitrary path $y(t)$ between the same endpoints is written as $y(t) = \bar{y}(t) + \eta(t)$, where $\eta(t)$ is any variation (not necessarily small) about $\bar{y}(t)$.

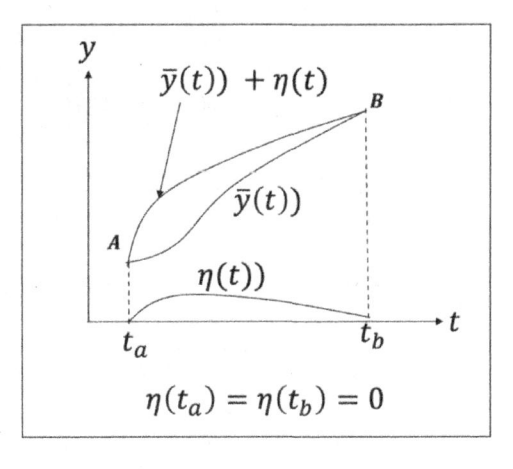

Since $y(t)$ and $\bar{y}(t)$ have the same values at $t = t_a$ and $t = t_b$, $\eta(t)$ must vanish at $t = t_a$ and $t = t_b$, i.e., $\eta(t_a) = \eta(t_b) = 0$.

Writing $y(t) = \bar{y}(t) + \eta(t)$ in Eq. (12), and expanding S about the least path $\bar{y}$, we get:

$$S = \int_{t=t_a}^{t=t_b} \left[\frac{m}{2}\left(\frac{d}{dt}(\bar{y} + \eta)\right)^2 + k(\bar{y} + \eta)\right] dt$$

$$= \int_{t=t_a}^{t=t_b} \left[\frac{m}{2}\left(\frac{d\bar{y}}{dt} + \frac{d\eta}{dt}\right)^2 + k(\bar{y} + \eta)\right] dt$$

Feynman's Path Integral explained with basic Calculus

$$= \int_{t=t_a}^{t=t_b} \left[\frac{m}{2}\left(\left(\frac{d\bar{y}}{dt}\right)^2 + 2\frac{d\bar{y}}{dt}\frac{d\eta}{dt} + \left(\frac{d\eta}{dt}\right)^2\right) + k(\bar{y}+\eta)\right]dt$$

[Using $(a+b)^2 = a^2 + 2ab + b^2$]

$$= \int_{t=t_a}^{t=t_b} \left[\left(\frac{m}{2}\left(\frac{d\bar{y}}{dt}\right)^2 + k\bar{y}\right) + m\frac{d\bar{y}}{dt}\frac{d\eta}{dt} + k\eta + \frac{m}{2}\left(\frac{d\eta}{dt}\right)^2\right]dt$$

[Regrouping the terms]

$$= \int_{t=t_a}^{t=t_b} \left[\frac{m}{2}\left(\frac{d\bar{y}}{dt}\right)^2 + k\bar{y}\right]dt + \int_{t=t_a}^{t=t_b} m\frac{d\bar{y}}{dt}\frac{d\eta}{dt}dt + \int_{t=t_a}^{t=t_b} \left[k\eta + \frac{m}{2}\left(\frac{d\eta}{dt}\right)^2\right]dt$$

The first integral in the above is: "Eq. (12) evaluated for the least path $y(t) = \bar{y}(t)$". Calling that integral S_0, we get from above:

$$S = S_0 + \int_{t=t_a}^{t=t_b} m\frac{d\bar{y}}{dt}\frac{d\eta}{dt}dt + \int_{t=t_a}^{t=t_b} \left[k\eta + \frac{m}{2}\left(\frac{d\eta}{dt}\right)^2\right]dt \quad \ldots\ldots\ldots \text{Eq. (14)}.$$

Integrating $\int_{t=t_a}^{t=t_b} \frac{d\bar{y}}{dt}\frac{d\eta}{dt}dt$ in Eq. (14) by parts, we obtain:

$$\int_{t=t_a}^{t=t_b} \frac{d\bar{y}}{dt}\frac{d\eta}{dt}dt = \frac{d\bar{y}}{dt}\eta(t)\Big|_{t=t_a}^{t=t_b} - \int_{t=t_a}^{t=t_b} \eta(t)\frac{d}{dt}\left(\frac{d\bar{y}}{dt}\right)dt$$

Now, since $\eta(t_b) = \eta(t_a) = 0$, the first term on the right of the above equation vanishes, and we are left with:

$$\int_{t=t_a}^{t=t_b} \frac{d\bar{y}}{dt}\frac{d\eta}{dt}dt = -\int_{t=t_a}^{t=t_b} \eta(t)\frac{d^2\bar{y}}{dt^2}dt, \text{ using which in Eq. (14), we obtain:}$$

$$S = S_0 - m\int_{t=t_a}^{t=t_b} \eta(t)\frac{d^2\bar{y}}{dt^2}dt + \int_{t=t_a}^{t=t_b} k\eta\, dt + \int_{t=t_a}^{t=t_b} \frac{m}{2}\left(\frac{d\eta}{dt}\right)^2 dt$$

$$= S_0 - \int_{t=t_a}^{t=t_b} dt\,\eta(t)\left[m\frac{d^2\bar{y}}{dt^2} - k\right] + \int_{t=t_a}^{t=t_b} \frac{m}{2}\left(\frac{d\eta}{dt}\right)^2 dt \quad \ldots\ldots \text{Eq. (15)}.$$

Since, in Eq. (15), S_0 is the action for the least path, S is equal to S_0 up to the first order variation in η, per the stationarity principle (Chapter 2). Hence, in Eq. (15), momentarily ignoring the third term $\int_{t=t_a}^{t=t_b} \frac{m}{2}\left(\frac{d\eta}{dt}\right)^2 dt$ (a second order term in the derivative of η), and setting S to S_0, we get $\int_{t=t_a}^{t=t_b} \eta(t)\left[m\frac{d^2\bar{y}}{dt^2} - k\right]dt = 0$, implying $m\frac{d^2\bar{y}}{dt^2} - k = 0$, since $\eta(t)$ is arbitrary. Plugging $m\frac{d^2\bar{y}}{dt^2} - k = 0$ back in Eq. (15), we get:

Chapter 5: Path Integrals in the Presence of Potential Energies

$$S = S_0 + \int_{t=t_a}^{t=t_b} \frac{m}{2}\left(\frac{d\eta}{dt}\right)^2 dt \quad \text{............................ Eq. (16).}$$

Note that the square of the derivative of η, although a higher order term, is very much present in the final expression of the action S (Eq. (16)). It is only for the sake of applying the stationarity principle that we had momentarily ignored it.

Terms linear in η are absent from the action in Eq. (16), thanks to η being the variation about the least path. [See Figure 18 (in Chapter 3) and the associated discussion on the absence of a linear term in the action.]

Notice how in Eq. (16), S_0, the action of the least path $\bar{y}(t)$, is completely separated from an integral involving $\eta(t)$. Denoting the integral $\int_{t=t_a}^{t=t_b} \frac{m}{2}\left(\frac{d\eta}{dt}\right)^2 dt$ in Eq. (16) as S_η, we have:

$$S = S_0 + S_\eta \quad \text{... Eq. (17).}$$

Next, we replace the y variables in the Path integral of Eq. (13) by the corresponding η variables as follows. $y(t) = \bar{y}(t) + \eta(t)$. Hence:

$$y_1 \equiv y(t = t_1)$$
$$= \bar{y}(t = t_1) + \eta(t = t_1)$$
$$= \bar{y}_1 + \eta_1 \quad [\text{We used } \bar{y}_1 \equiv \bar{y}(t = t_1)]$$

So, $y_1 = \bar{y}_1 + \eta_1$. Similarly, $y_2 \equiv y(t = t_2) = \bar{y}_2 + \eta_2$,.....and so on.

Considering $y_1 = \bar{y}_1 + \eta_1$, we have, $dy_1 = d\bar{y}_1 + d\eta_1$. But $d\bar{y}_1 = 0$, since $\bar{y}_1$, being on the least path, is "fixed" and hence, a constant. So, $dy_1 = d\eta_1$. Similarly, $dy_2 = d\eta_2$, and so on. We write Eq. (13) in terms of the η variables as follows:

$$K(t_b, y_b; t_a, y_a)$$
$$= \left(\frac{1}{A}\right)^N \int_{-\infty}^{+\infty} d\eta_{N-1} \int_{-\infty}^{+\infty} d\eta_{N-2} \ldots \ldots \int_{-\infty}^{+\infty} d\eta_3 \int_{-\infty}^{+\infty} d\eta_2 \int_{-\infty}^{+\infty} d\eta_1 \, e^{\frac{i}{\hbar}S}$$
$$\text{................................ Eq. (18).}$$

Feynman's Path Integral explained with basic Calculus

Next, with the help of Eq. (17), we can write $e^{\frac{i}{\hbar}S}$ in the integrand of Eq. (18) as: $e^{\frac{i}{\hbar}S} = e^{\frac{i}{\hbar}(S_0 + S_\eta)} = e^{\frac{i}{\hbar}S_0} e^{\frac{i}{\hbar}S_\eta}$, using which in Eq. (18) we obtain:

$$K(t_b, y_b; t_a, y_a)$$
$$= \left(\frac{1}{A}\right)^N \int_{-\infty}^{+\infty} d\eta_{N-1} \int_{-\infty}^{+\infty} d\eta_{N-2} \ldots \int_{-\infty}^{+\infty} d\eta_3 \int_{-\infty}^{+\infty} d\eta_2 \int_{-\infty}^{+\infty} d\eta_1 \, e^{\frac{i}{\hbar}S_0} e^{\frac{i}{\hbar}S_\eta}$$

Since, $e^{\frac{i}{\hbar}S_0}$ does not depend on the η's, we can take it outside the integral. Hence, from the above:

$$K(t_b, y_b; t_a, y_a)$$
$$= e^{\frac{i}{\hbar}S_0} \int_{-\infty}^{+\infty} d\eta_{N-1} \ldots \int_{-\infty}^{+\infty} d\eta_2 \int_{-\infty}^{+\infty} d\eta_1 \left(\frac{1}{A}\right)^N e^{\frac{i}{\hbar}S_\eta} \quad \ldots\ldots\ldots\ldots \text{Eq. (19).}$$

In Eq. (19), we will leave the multiple integral $\left(\frac{1}{A}\right)^N \int_{-\infty}^{+\infty} d\eta_{N-1} \ldots \int_{-\infty}^{+\infty} d\eta_1 \, e^{\frac{i}{\hbar}S_\eta}$, which depends only on the η's, unevaluated, treating the integral as a constant that multiplies $e^{\frac{i}{\hbar}S_0}$.

Hence, from Eq. (19), the desired Path Integral is:

$$K(t_b, y_b; t_a, y_a) = e^{\frac{i}{\hbar}S_0} (a \ constant) \quad \ldots\ldots\ldots\ldots\ldots\ldots \text{Eq. (20).}$$

Eq. (20) gives the propagator for a particle having a potential energy that varies linearly with the position or the space variable. As we mentioned in the section "Indirect Method for Evaluating Path Integrals" in Chapter 3, the constant such as the one in Eq. (20), is usually evaluated from some other consideration, which we won't worry about here. S_0 is the action in Eq. (12) evaluated for the least path or classical path $\bar{y}(t)$. Note how the action for the classical path plays a crucial role in quantum mechanics. As for S_0 in Eq. (20), you do not need to evaluate S_0 from the scratch; you can borrow the result from the exercise associated with Example 2 from Chapter 2, which essentially deals with the same action as in this problem, like we mentioned a while ago. You can use S_0 obtained in Eq. (29) of Chapter 2; all you have to do is replace

Chapter 5: Path Integrals in the Presence of Potential Energies

mg in that result by k. Carrying out the above steps, we obtain from Eq. (29) of Chapter 2:

$$S_0 = \frac{k^2}{3m}(t_b^3 - t_a^3) + k(t_b^2 - t_a^2)C_1 + \left(\frac{1}{2}mC_1^2 + kC_2\right)(t_b - t_a)$$

$$\dots\dots\dots\dots\dots\dots\dots \text{Eq. (21).}$$

where $C_1 = \frac{y_b - y_a}{t_b - t_a} - \frac{1}{2}\frac{k}{m}(t_a + t_b)$, and $C_2 = \frac{y_a t_b - y_b t_a}{t_b - t_a} + \frac{1}{2}\frac{k}{m}t_b t_a$.

Hence, per Eq. (20), the propagator $K(t_b, x_b; t_a, x_a)$ for this problem is $(constant)e^{\frac{i}{\hbar}S_0}$, where S_0 is given by Eq. (21).

Note that if you set $k = 0$ in this problem, you have no potential energy (since the potential energy $-ky$ is zero when $k = 0$). Hence, S_0, as given by Eq. (21), should reduce to the free-particle-action of the corresponding classical path or least path, and the propagator given by Eq. (20) should reduce to the free particle propagator. Let us check if that is the case. Using $k = 0$ in S_0, as given by Eq. (21), we get $S_0 = \frac{1}{2}mC_1^2(t_b - t_a) = \frac{1}{2}m\frac{(y_b - y_a)^2}{t_b - t_a}$ (We used $C_1 = \frac{y_b - y_a}{t_b - t_a}$ that we obtained by setting $k = 0$ in the expression for C_1).

Now, (as anticipated), $S_0 = \frac{1}{2}m\frac{(y_b - y_a)^2}{t_b - t_a}$ is precisely the free-particle-action for the classical path, using which in Eq. (20), we get $K(t_b, y_b; t_a, y_a) = e^{\frac{im(y_b - y_a)^2}{2\hbar(t_b - t_a)}}(a\ constant)$, the same expression as obtained in Eq. (5) from Chapter 4 for the free particle propagator, except for an undetermined constant factor. (You cannot determine the constant factor using the indirect method.)

Example 2 (The Potential Energy is Quadratic in the Position Variable)
Calculate the propagator for the action with a potential energy varying quadratically with the position or the space variable: $S = \int_{t=t_a}^{t=t_b}\left[\frac{m}{2}\left(\frac{dx(t)}{dt}\right)^2 - \frac{1}{2}kx^2\right]dt$, by using the indirect method.

Feynman's Path Integral explained with basic Calculus

By this time, you have probably realized that you need to express an arbitrary path $x(t)$ between two fixed points as: $x(t) = \bar{x}(t) + \eta(t)$, where $\bar{x}(t)$ is the "the least or the classical path" for a given action and $\eta(t)$ is an arbitrary variation (with fixed endpoints) about the least path ($\eta(t_b) = \eta(t_a) = 0$). If you can write $S = S_0 + S_\eta$, where S_0 is action for $\bar{x}(t)$ and S_η depends entirely on the variable $\eta(t)$, then the propagator $K(t_b, x_b; t_a, x_a)$ is simply given as $e^{\frac{i}{\hbar}S_0}$ (a constant). [If above argument is not clear, go back to the previous problem (Example 1) and review the indirect method for Path Integrals.]

Continuing, S (as given in the problem statement) for an arbitrary path $x(t)$ is expanded about the least path $\bar{x}(t)$ as follows:

$$S = \int_{t=t_a}^{t=t_b} \left[\frac{m}{2}\left(\frac{d}{dt}(\bar{x}+\eta)\right)^2 - \frac{1}{2}k(\bar{x}+\eta)^2\right]dt$$

$$= \int_{t=t_a}^{t=t_b} \left[\frac{m}{2}\left(\frac{d\bar{x}}{dt}+\frac{d\eta}{dt}\right)^2 - \frac{1}{2}k(\bar{x}+\eta)^2\right]dt$$

$$= \int_{t=t_a}^{t=t_b} \left[\frac{m}{2}\left(\left(\frac{d\bar{x}}{dt}\right)^2 + 2\frac{d\bar{x}}{dt}\frac{d\eta}{dt} + \left(\frac{d\eta}{dt}\right)^2\right) - \frac{1}{2}k(\bar{x}^2 + 2\bar{x}\eta + \eta^2)\right]dt$$

[Using $(a+b)^2 = a^2 + 2ab + b^2$]

Regrouping the terms in the expression above, we get:

$$S = \int_{t=t_a}^{t=t_b}\left(\frac{m}{2}\left(\frac{d\bar{x}}{dt}\right)^2 - \frac{1}{2}k\bar{x}^2\right)dt + \int_{t=t_a}^{t=t_b}\left(m\frac{d\bar{x}}{dt}\frac{d\eta}{dt} - k\bar{x}\eta\right)dt +$$

$$\int_{t=t_a}^{t=t_b}\left(\frac{m}{2}\left(\frac{d\eta}{dt}\right)^2 - \frac{1}{2}k\eta^2\right)dt \quad \ldots\ldots\ldots\ldots\ldots\ldots \text{Eq. (22)}.$$

The first integral in Eq. (22) is the action $S = \int_{t=t_a}^{t=t_b}\left[\frac{m}{2}\left(\frac{dx(t)}{dt}\right)^2 - \frac{1}{2}kx^2\right]dt$ evaluated for the least or the classical path $x(t) = \bar{x}$. Calling the action for the least path S_0, we get from Eq. (22):

$$S = S_0 + \int_{t=t_a}^{t=t_b}\left[m\frac{d\bar{x}}{dt}\frac{d\eta}{dt} - k\bar{x}\eta\right] + \int_{t=t_a}^{t=t_b}\left(\frac{m}{2}\left(\frac{d\eta}{dt}\right)^2 - \frac{1}{2}k\eta^2\right)dt$$

$$\ldots\ldots\ldots\ldots\ldots\ldots \text{Eq. (23)}.$$

Chapter 5: Path Integrals in the Presence of Potential Energies

Integrating $\int_{t=t_a}^{t=t_b} \frac{d\bar{x}}{dt} \frac{d\eta}{dt} dt$ appearing on the right of Eq. (23) by parts, we obtain:

$$\int_{t=t_a}^{t=t_b} \frac{d\bar{x}}{dt} \frac{d\eta}{dt} dt = \frac{d\bar{x}}{dt} \eta(t) \Big|_{t=t_a}^{t=t_b} - \int_{t=t_a}^{t=t_b} \eta(t) \frac{d}{dt}\left(\frac{d\bar{x}}{dt}\right) dt$$

Now, since $\eta(t_b) = \eta(t_a) = 0$, the first term on the right of the above equation vanishes, and we are left with:

$$\int_{t=t_a}^{t=t_b} \frac{d\bar{x}}{dt} \frac{d\eta}{dt} dt = -\int_{t=t_a}^{t=t_b} \eta(t) \frac{d^2\bar{x}}{dt^2} dt,$$ using which in Eq. (23), we obtain:

$$S = S_0 - \int_{t=t_a}^{t=t_b} dt\, \eta(t) \left[m\frac{d^2\bar{x}}{dt^2} + k\bar{x}\right] + \int_{t=t_a}^{t=t_b} \left[\frac{m}{2}\left(\frac{d\eta}{dt}\right)^2 - \frac{1}{2}k\eta^2\right] dt$$

.... Eq. (24).

Now, since S_0 and S in Eq. (24) correspond to the least path and a variation about the least path respectively, per stationarity principle (Chapter 2), $S = S_0$ upto the first order variation $\eta(t)$. So, momentarily ignoring the second integral (since it consists of second order terms) in Eq. (24) and setting $S = S_0$, we get, $\int_{t=t_a}^{t=t_b} dt\, \eta(t) \left[m\frac{d^2\bar{x}}{dt^2} + k\bar{x}\right] = 0$. Since $\eta(t)$ is arbitrary, we have $m\frac{d^2\bar{x}}{dt^2} + k\bar{x} = 0$. Plugging $m\frac{d^2\bar{x}}{dt^2} + k\bar{x} = 0$ back in Eq. (24), we obtain:

$$S = S_0 + \int_{t=t_a}^{t=t_b} \left[\frac{m}{2}\left(\frac{d\eta}{dt}\right)^2 - \frac{1}{2}k\eta^2\right] dt.$$

Note that in the above expression, S_0, the action corresponding to the least path $\bar{x}(t)$, is separated from an integral involving only $\eta(t)$, just what we were hoping for. We will denote the second term, i.e., $\int_{t=t_a}^{t=t_b} \left[\frac{m}{2}\left(\frac{d\eta}{dt}\right)^2 - \frac{1}{2}k\eta^2\right] dt$, as S_η. Hence, we have $S = S_0 + S_\eta$. Per our discussion at the beginning of the problem, this implies that the probability amplitude $K(t_b, x_b; t_a, x_a)$ for a particle with a quadratic PE, going from A (t_a, x_a) to B(t_b, x_b), is given as $e^{\frac{i}{\hbar}S_0}(a\ constant)$: $K(t_b, x_b; t_a, x_a) = e^{\frac{i}{\hbar}S_0}(a\ constant)$. In Exercise 4 from

96

Feynman's Path Integral explained with basic Calculus

Chapter 2, we evaluated the action S_0 for the classical path or least path for this problem, by using the initial conditions $t_a = 0$ and $x_a = 0$, and obtained

$$S_0 = \frac{m\omega x_b^2 \cos \omega t_b}{2 \sin \omega t_b}$$ (Eq. (37) from Chapter 2), where $\omega = \sqrt{\frac{k}{m}}$. We will use this S_0 in $K(t_b, x_b; t_a, x_a)$ with $t_a = 0, x_a = 0$ and obtain:

$$K(t_b, x_b; t_a = 0, x_a = 0) = (constant) e^{\frac{i m \omega x_b^2 \cos \omega t_b}{\hbar \; 2 \sin \omega t_b}} \quad \ldots\ldots\text{Eq. (25)},$$

which gives the probability amplitude for a particle with a potential energy quadratic in the position variable, to go from $(0, 0)$ to (t_b, x_b).

When $k = 0$, the potential energy in this problem, i.e., $\frac{1}{2}kx^2$ is zero; hence, in the $k = 0$ limit, Eq. (25) should reduce to the free particle propagator (up to a constant). Let us check if that is the case.

When $k = 0$, $\omega = \sqrt{\frac{k}{m}}$ is also zero. We will obtain expressions for $\sin \omega t_b$ and $\cos \omega t_b$ in the $\omega \to 0$ limit and use them in Eq. (25). Using the Taylor expansions of the sines and the cosine, as given by Eq. (12) and Eq. (13) of Chapter 1, and keeping only the highest order terms, we get

$\sin \omega t_b \approx \omega t_b$, and $\cos \omega t_b \approx 1$, using which in Eq. (25), we get after some algebra:

$$K(t_b, x_b; t_a = 0, x_a = 0) = (constant) e^{\frac{i m x_b^2}{2 \hbar t_b}}$$

The above expression agrees with the free particle propagator given by Eq. (5) of Chapter 4 (with $t_a = 0, x_a = 0$), apart from a constant factor.

In the last two chapters, we saw how the classical path plays an important role even in quantum mechanics, via the propagator.

Chapter 5: Path Integrals in the Presence of Potential Energies

RECAP: <u>Indirect Method for Evaluating a Path Integral</u>

- Express an arbitrary path $x(t)$ between two fixed points as: $x(t) = \bar{x}(t) + \eta(t)$, where $\bar{x}(t)$ is the "the least or classical path" for a given action S and $\eta(t)$ is an arbitrary variation (with fixed endpoints) about the classical path.

- If $S = S_0 + S_\eta$, where S_0 is action for the least path $\bar{x}(t)$ and S_η depends entirely on the variable $\eta(t)$ and its derivatives, then the propagator is given by:

$$K(t_b, x_b; t_a, x_a) = e^{\frac{i}{\hbar}S_0}(a\ constant)$$

Feynman's Path Integral explained with basic Calculus

Chapter 6: Deriving Schrodinger's Equation Using the Propagator

In this section, we will derive Schrodinger's equation using the propagator. For that, we first need to introduce the concept of a wavefunction. We know that a propagator gives the probability amplitude for a particle to go from one point to another. A wavefunction, on the other hand, gives the probability amplitude for a particle to be at a specific space-time point. A wavefunction is similar to a propagator in a sense, but unlike a propagator, a wavefunction does not have any information about where the particle came from. Sometimes all we care about is the probability amplitude $\psi(t,x)$ for a particle to be at the space-time point (t,x). Because, if we know $\psi(t,x)$, we can take its absolute square $|\psi(t,x)|^2$ and call that the probability density $P(t,x)$ of finding the particle at the position x, at time t.

Our goal is to derive a differential equation for $\psi(t,x)$, i.e., Schrodinger's equation. To that end, we will use a formula similar to $K(B,A) = \int_{-\infty}^{+\infty} dx_c K(B,C) K(C,A)$ (Eq. (6) from Chapter 5.) We will replace the propagators $K(C,A)$ in the integrand and $K(B,A)$ on the left of the equation (each of which has A as the initial point), by the wavefunctions $\psi(C)$ and $\psi(B)$ respectively, since we do not care about the initial point A. Hence, Eq. (6) in Chapter 5 assumes the following form:

$$\psi(B) = \int_{-\infty}^{+\infty} dx_c K(B,C) \psi(C) \quad \text{............................ Eq. (1).}$$

where $K(B,C)$ is the only propagator and the rest are wavefunctions.

Feynman's Path Integral explained with basic Calculus

Let, C be the space-time point (t, y) (Figure 31). y is just a label for the space variable x at time t and will play the role of x_c in Eq. (1). The wavefunction at (t, y) is $\psi(t, y)$. B is the space-time point $(t + \epsilon, x)$, where $t + \epsilon$ is only slightly later than t. We are interested in the wavefunction $\psi(B) \equiv \psi(t + \epsilon, x)$ at $t + \epsilon$. The propagator $K(B, C) \equiv K(t + \epsilon, x; t, y)$ takes the wavefunction from the space-time point (t, y) to the space-time point $(t + \epsilon, x)$. Summarizing, in Eq. (1), we use:

$$\psi(C) = \psi(t, y)$$
$$K(B, C) = K(t + \epsilon, x; t, y)$$
$$\psi(B) = \psi(t + \epsilon, x)$$

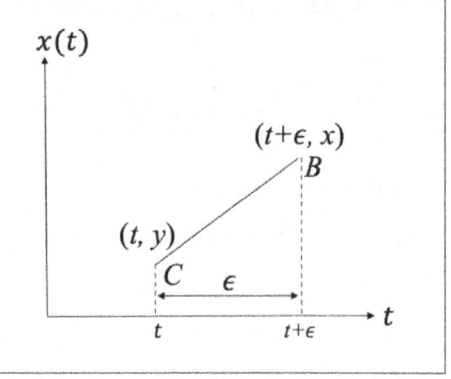

Figure 31

We obtain:

$$\psi(t + \epsilon, x) = \int_{-\infty}^{+\infty} dy\, K(t + \epsilon, x; t, y) \psi(t, y) \quad \text{Eq. (2).}$$

(Note that y in Eq. (2) is the integration variable). We will consider different propagators in Eq. (2), depending on whether we are dealing with a free particle or a particle with potential energy. But before getting into that, we revisit Taylor expansion that we encountered in Chapter 1, but this time, involving two variables. In our two-variable Taylor expansion, one of the two variables will be kept fixed. For example, while expanding $f(t + \epsilon, x)$ about the point (t, x), we will treat x as a constant. (t is a variable, its value changes from t to $t + \epsilon$). The rest is just like the Taylor expansion involving a single variable, with the novelty that instead of $\frac{d}{dt}$, we use $\frac{\partial}{\partial t}$ (partial derivative), the

Chapter 6: Deriving Schrodinger's Equation using the Propagator

symbol $\frac{\partial}{\partial t}$ indicating that the other variable (x) is maintained at a constant value. Hence, following Eq. (11) of Chapter 1,

$$f(t + \epsilon, x) = f(t, x) + \epsilon \frac{\partial f}{\partial t} + \frac{\epsilon^2}{2!} \frac{\partial^2 f}{\partial t^2} + \cdots \qquad [x \text{ is fixed}].$$

t in the above is just like x in Eq. (11) of Chapter 1, whereas x is treated a constant; ϵ is like $(x - a)$ in Eq. (11) of Chapter 1. If you keep the terms up to only the term linear in ϵ (also referred as $o(\epsilon)$) in the above series expansion, you get:

$$f(t + \epsilon, x) \approx f(x, t) + \epsilon \frac{\partial f}{\partial t} \dots\dots\dots\dots\dots\dots\dots\dots\dots\dots \text{Eq. (3)}.$$

Eq. (3) is the $o(\epsilon)$ expansion of $f(x, t)$ in time t.

Similarly, you can keep t fixed and Taylor expand in the x-variable following Eq. (11) of Chapter 1:

$$f(t, x + \eta) = f(t, x) + \eta \frac{\partial f}{\partial x} + \frac{\eta^2}{2!} \frac{\partial^2 f}{\partial x^2} + \cdots \qquad [t \text{ is fixed}]$$

$$\dots\dots\dots\dots\dots\dots\dots\dots\dots \text{Eq. (4)}.$$

In the above equation, $\frac{\partial}{\partial x}$ is just like $\frac{d}{dx}$, with t treated as a constant. η in Eq. (4) is the same as $(x - a)$ in Eq. (11) of Chapter 1.

RECAP: Partial versus Ordinary Derivatives

Let, $f(x, t)$ be a function of the variables x and t

- $\frac{\partial}{\partial t} f(x, t)$ is the ordinary derivative of f with respect to t, when x is kept fixed.

- $\frac{\partial}{\partial x} f(x, t)$ is the ordinary derivative of f with respect to x, when t is kept fixed.

Feynman's Path Integral explained with basic Calculus

Example-1: Deriving Schrodinger's Equation for a Free Particle

In this example, using the free particle propagator for $K(t + \epsilon, x; t, y)$ on the right of Eq. (2), we will obtain Schrodinger's equation for a free particle. In Eq. (5) of Chapter 4, the free particle propagator was given as:

$$K(t_b, x_b; t_a, x_a) = \sqrt{\frac{m}{2\pi i\hbar(t_b-t_a)}} e^{\frac{im(x_b-x_a)^2}{2\hbar(t_b-t_a)}}.$$ In it, setting $t_b = t + \epsilon$, $t_a = t$, and $x_b = x$, $x_a = y$, we get:

$$K(t + \epsilon, x; t, y) = \sqrt{\frac{m}{2\pi i\hbar\epsilon}} e^{\frac{im(x-y)^2}{2\hbar\epsilon}},$$ using which in Eq. (2) (of the current chapter), we obtain:

$$\psi(t + \epsilon, x) = \sqrt{\frac{m}{2\pi i\hbar\epsilon}} \int_{-\infty}^{+\infty} dy\, e^{\frac{im(x-y)^2}{2\hbar\epsilon}} \psi(t, y) \quad \text{......... Eq. (5).}$$

In the above integral, the integration variable is y, and x is a constant. We define a new variable $\eta \equiv y - x$, implying $y = \eta + x$ and $d\eta = dy$ (since x is a constant). The limits of integration for this new variable η is from $-\infty$ to $+\infty$ just like the old variable y. Hence, in terms of η, the integral in Eq. (5) assumes the following form:

$$\psi(t + \epsilon, x) = \sqrt{\frac{m}{2\pi i\hbar\epsilon}} \int_{-\infty}^{+\infty} d\eta\, e^{\frac{im\eta^2}{2\hbar\epsilon}} \psi(t, x + \eta) \quad \text{......... Eq. (6).}$$

We Taylor-expand the left side of Eq. (6), by following Eq. (3):

$$\psi(t + \epsilon, x) = \psi(t, x) + \epsilon \frac{\partial \psi}{\partial t} + \cdots \quad \text{......... Eq. (7).}$$

In Eq. (7), we expanded up to $o(\epsilon)$. We also Taylor-expand $\psi(t, x + \eta)$ inside the integral on the right of Eq. (6), by following Eq. (4). We obtain:

$$\psi(t, x + \eta) = \psi(t, x) + \eta \frac{\partial \psi}{\partial x} + \frac{\eta^2}{2} \frac{\partial^2 \psi}{\partial x^2} + \cdots \quad \text{......... Eq. (8).}$$

In Eq. (8), we kept terms up to the second order in η, since, as you will see later, the η^2-term has an $o(\epsilon)$ contribution when integrated. More clearly

Chapter 6: Deriving Schrodinger's Equation using the Propagator

speaking, the η^2-term will contribute a term linear in ϵ, after being integrated in Eq. (6).

Using Eq. (7) and Eq. (8) in Eq. (6), we get:

$$\psi(t,x) + \epsilon \frac{\partial \psi}{\partial t}$$

$$= \sqrt{\frac{m}{2\pi i \hbar \epsilon}} \int_{-\infty}^{+\infty} d\eta \, e^{\frac{im\eta^2}{2\hbar\epsilon}} \left[\psi(t,x) + \eta \frac{\partial \psi}{\partial x} + \frac{\eta^2}{2} \frac{\partial^2 \psi}{\partial x^2} + \cdots \right]$$

$$= \sqrt{\frac{m}{2\pi i \hbar \epsilon}} \int_{-\infty}^{+\infty} d\eta \, e^{\frac{im\eta^2}{2\hbar\epsilon}} \psi(t,x) +$$

$$\sqrt{\frac{m}{2\pi i \hbar \epsilon}} \int_{-\infty}^{+\infty} d\eta \, e^{\frac{im\eta^2}{2\hbar\epsilon}} \eta \frac{\partial \psi}{\partial x} + \sqrt{\frac{m}{2\pi i \hbar \epsilon}} \int_{-\infty}^{+\infty} d\eta \, e^{\frac{im\eta^2}{2\hbar\epsilon}} \frac{\eta^2}{2} \frac{\partial^2 \psi}{\partial x^2} + \cdots$$

$$\ldots\ldots\ldots\ldots\ldots\ldots\ldots\ldots \text{Eq. (9)}.$$

In the above equation, $\psi(t,x)$ and its derivatives are treated as constants, since η is the integration variable. The second integral on the right of Eq. (9) is zero, since the integrand of the second integral is an odd function of η: In the integrand, $e^{\frac{im\eta^2}{2\hbar\epsilon}}$ is even in η, and η is odd in η. Hence, the product $\eta e^{\frac{im\eta^2}{2\hbar\epsilon}}$ is odd in η.

To evaluate the first and the third integrals in Eq. (9), we will use results from Chapter 1. The first integral in Eq. (9) is evaluated by using Eq. (2) of Chapter (1) with $a = 0$, viz., $\int_{-\infty}^{+\infty} d\eta \, e^{-\frac{\eta^2}{2\sigma^2}} = \sigma\sqrt{2\pi}$. For our current problem, we set $-\frac{1}{2\sigma^2} = \frac{im}{2\hbar\epsilon}$, i.e., $\sigma = \sqrt{\frac{i\hbar\epsilon}{m}}$ in the integral to obtain:

$$\int_{-\infty}^{+\infty} d\eta \, e^{\frac{im\eta^2}{2\hbar\epsilon}} = \sqrt{\frac{2\pi i \hbar \epsilon}{m}} \ldots\ldots\ldots\ldots\ldots\ldots\ldots \text{Eq. (10)}.$$

Using Eq. (10) in the first term on the right of Eq. (9), we obtain $\psi(x,t)$ for the first term.

As for the integral in the third term on the right of Eq. (9), we will use Eq. (18) from Chapter 1, viz.,

Feynman's Path Integral explained with basic Calculus

$$\frac{1}{\sigma\sqrt{2\pi}} \int_{-\infty}^{+\infty} d\eta\, \eta^2 e^{-\frac{\eta^2}{2\sigma^2}} = \sigma^2$$

By setting $\sigma = \sqrt{\frac{i\hbar\epsilon}{m}}$ in the above integral (This is the same σ that was used in the derivation of Eq. (10)), we get:

$$\int_{-\infty}^{+\infty} d\eta\, e^{\frac{im\eta^2}{2\hbar\epsilon}} \eta^2 = \left(\frac{i\hbar\epsilon}{m}\right)^{\frac{3}{2}} \sqrt{2\pi} \quad\dots\dots\dots\dots\dots \text{Eq. (11).}$$

Using Eq. (11) in the third term on the right of Eq. (9), the third term is reduced to $\frac{i\hbar\epsilon}{2m}\frac{\partial^2 \psi}{\partial x^2}$, which is first order or linear in the time variable ϵ: $o(\epsilon)$. Hence, an integral involving the second order space variable η produces a term that is first order or linear in the time variable ϵ: $o(\epsilon)$.

Summarizing, we have $\psi(x,t)$ and $\frac{i\hbar\epsilon}{2m}\frac{\partial^2 \psi}{\partial x^2}$ for the first and the third terms on the right side of Eq. (9) respectively, and the second term on the right side of Eq. (9) is zero. The rest of the terms are all $o(\epsilon^2)$ or higher. Hence, keeping only terms up to those linear in ϵ, on both sides of Eq. (9), we obtain:

$$\psi(x,t) + \epsilon \frac{\partial \psi}{\partial t} = \psi(x,t) + \epsilon \frac{i\hbar}{2m}\frac{\partial^2 \psi}{\partial x^2}$$

$$\Rightarrow \epsilon \frac{\partial \psi}{\partial t} = \epsilon \frac{i\hbar}{2m}\frac{\partial^2 \psi}{\partial x^2}$$

Cancelling ϵ from both sides and multiplying both sides by $i\hbar$, we get

$$i\hbar \frac{\partial \psi}{\partial t} = -\frac{\hbar^2}{2m}\frac{\partial^2 \psi}{\partial x^2} \quad\dots\dots\dots\dots\dots \text{Eq. (12).}$$

Eq. (12) is Schrodinger's equation for a free particle.

RECAP: Two Key Integrals for deriving Schrodinger's Equation

1. $\int_{-\infty}^{+\infty} d\eta\, e^{-\frac{\eta^2}{2\sigma^2}} = \sigma\sqrt{2\pi}$, and

2. $\frac{1}{\sigma\sqrt{2\pi}} \int_{-\infty}^{+\infty} d\eta\, \eta^2 e^{-\frac{\eta^2}{2\sigma^2}} = \sigma^2$

Chapter 6: Deriving Schrodinger's Equation using the Propagator

Example-2: Deriving Schrodinger's Equation for a particle having a potential energy:

Next, we will find Schrodinger's equation for a particle having a potential energy. Once again, we will use Eq. (2); but what to use for $K(t + \epsilon, x; t, y)$? In the last problem, we used the complete expression for the free particle propagator for K valid for any time interval. But we did not need to; here is why. Since, in Eq. (2), we are dealing with a propagator defined over a small time-interval, we could have used an infinitesimal-time free particle propagator, as given by Eq. (9) of Chapter 4, in Eq. (2) of this chapter, and still got the same answer as Eq. (12). That derivation would have been similar to the following treatment of the "particle with potential energy" case, in which, we will use the infinitesimal-time propagator with the potential energy (PE), as given by Eq. (4) of Chapter 5, viz.,

$$K(t_{i+1}, x_{i+1}; t_i, x_i) = \frac{1}{A} e^{\frac{i}{\hbar}\left[\frac{m(x_{i+1}-x_i)^2}{2\epsilon} - \epsilon V\left(\frac{x_{i+1}+x_i}{2}\right)\right]},$$

where $t_{i+1} = t_i + \epsilon$. Borrowing notations from Example 1, the space-variable is called y for an "earlier" time and x for the "later" time. Also, the earlier time t_i will be simply denoted as t and the later time t_{i+1} as $t + \epsilon$. Hence, $t_i = t$, $x_i = y$, and $t_{i+1} = t + \epsilon$, $x_{i+1} = x$. Writing the above infinitesimal-time propagator in terms of these variables, we get:

$$K(t + \epsilon, x; t, y) = \frac{1}{A} e^{\frac{i}{\hbar}\left[\frac{m(x-y)^2}{2\epsilon} - \epsilon V\left(\frac{x+y}{2}\right)\right]} \quad\quad\quad\quad\quad \text{Eq. (13).}$$

You may be wondering what A to use in Eq. (13). Once we substitute $K(t + \epsilon, x; t, y)$ in Eq. (2) by Eq. (13), we will determine a definite value of A. As you will see, we will compare the two sides of Eq. (2), after Taylor-expanding the terms on both sides and carrying out integrals like those in Example 1. You will also see that despite the presence of the potential energy

Feynman's Path Integral explained with basic Calculus

V, you get the same A as in the free particle case, just like we said you would previously. Note that we never formally derived the expression for $\frac{1}{A}$ for a free particle (Refer to Eq. (9) from Chapter 4). The constant factor that multiplies the complex exponential function in the free particle propagator was not formally derived. In Eq. (5) from Chapter 4, we guessed its form drawing inspiration from another problem. That will be remedied as we derive $\frac{1}{A}$ for a particle with a potential energy in the following, since you can extend the method effortlessly to the case of the free particle.

As in the previous chapters, we will focus on two different potential energies, viz., linear and quadratic in the space variable, and derive Schrodinger's equation for them using Eq. (2) and Eq. (13). Finally, we will derive Schrodinger's equation for a general potential, by using the same approach.

Case I: The Potential Energy $V(x)$ Is Linear in the Space Variable

In this problem, we consider the potential energy $V(x)$ having a linear dependence on the space variable x: $V(x) = -kx$. Hence $V\left(\frac{x+y}{2}\right) = -\frac{1}{2}k(x+y)$, using which in Eq. (13), we get:

$$K(t+\epsilon, x; t, y) = \frac{1}{A} e^{\frac{i}{\hbar}\left[\frac{m(x-y)^2}{2\epsilon} + \epsilon\frac{1}{2}k(x+y)\right]} \quad \text{.....................Eq. (14).}$$

We will use Eq. (14) in Eq. (2), viz., $\psi(t+\epsilon, x) = \int_{-\infty}^{+\infty} dy K(t+\epsilon, x; t, y)\psi(t, y)$, to get:

$$\psi(t+\epsilon, x) = \int_{-\infty}^{+\infty} \frac{1}{A} dy e^{\frac{i}{\hbar}\left[\frac{m(x-y)^2}{2\epsilon} + \epsilon k\left(\frac{x+y}{2}\right)\right]} \psi(t, y)$$

$$= \frac{1}{A} \int_{-\infty}^{+\infty} dy e^{\frac{im(x-y)^2}{\hbar 2\epsilon}} e^{\frac{i}{\hbar}\epsilon k\left(\frac{x+y}{2}\right)} \psi(t, y)$$

$$\text{............................ Eq. (15).}$$

107

Chapter 6: Deriving Schrodinger's Equation using the Propagator

Defining $\eta \equiv y - x$, we have $d\eta = dy$ (Since x is fixed and hence a constant). η, like y, varies from $-\infty$ to $+\infty$. Hence, in terms of η, Eq. (15) assumes the following form:

$$\psi(t + \epsilon, x) = \frac{1}{A}\int_{-\infty}^{+\infty} d\eta\, e^{\frac{im\eta^2}{\hbar\, 2\epsilon}} e^{\frac{i}{\hbar}\epsilon k\left(x+\frac{\eta}{2}\right)} \psi(t, x + \eta) \quad \ldots\ldots\ldots \text{Eq. (16)}.$$

1. We Taylor-expand the left side of Eq. (16) as:

$$\psi(t + \epsilon, x) = \psi(t, x) + \epsilon\frac{\partial \psi}{\partial t} + \cdots$$

(By following Eq. (3))

2. We also Taylor-expand $\psi(t, x + \eta)$ in the integrand on the right of Eq. (16) as:

$$\psi(t, x + \eta) = \psi(t, x) + \eta\frac{\partial \psi}{\partial x} + \frac{\eta^2}{2}\frac{\partial^2 \psi}{\partial x^2} + \cdots$$

(By following Eq. (4))

3. In the integrand on the right side of Eq. (16), we Taylor-expand $e^{\frac{i}{\hbar}\epsilon k\left(x+\frac{\eta}{2}\right)}$ by using Eq. (10) of Chapter 1, viz. $e^u = 1 + u + \cdots$, and by setting $u = \frac{i}{\hbar}\epsilon k\left(x + \frac{\eta}{2}\right)$ as follows:

$$e^{\frac{i}{\hbar}\epsilon k\left(x+\frac{\eta}{2}\right)} = 1 + \frac{i}{\hbar}\epsilon k\left(x + \frac{\eta}{2}\right) + \cdots$$

(Keeping only terms up to $o(\epsilon)$)

Using 1, 2 and 3 in Eq. (16), we get:

$$\psi(t, x) + \epsilon\frac{\partial \psi}{\partial t} + \cdots$$

$$= \frac{1}{A}\int_{-\infty}^{+\infty} d\eta\, e^{\frac{im\eta^2}{\hbar\, 2\epsilon}} \left[1 + \frac{i}{\hbar}\epsilon kx + \frac{i}{\hbar}\epsilon k\frac{\eta}{2} + \cdots\right]$$

$$\left[\psi(t, x) + \eta\frac{\partial \psi}{\partial x} + \frac{\eta^2}{2}\frac{\partial^2 \psi}{\partial x^2} + \cdots\right]$$

$$= \frac{1}{A}\int_{-\infty}^{+\infty} d\eta\, e^{\frac{im\eta^2}{\hbar\, 2\epsilon}} \left[\psi(t, x) + \eta\frac{\partial \psi}{\partial x} + \frac{\eta^2}{2}\frac{\partial^2 \psi}{\partial x^2} + \frac{i}{\hbar}\epsilon kx\psi(t, x) + \cdots\right]$$

$$\ldots\ldots\ldots\ldots\ldots\ldots \text{Eq. (17)}.$$

While evaluating various terms in the integral in Eq. (17), we need to be aware of the following:

Feynman's Path Integral explained with basic Calculus

1. $x, \psi(t,x), \frac{\partial \psi}{\partial x}, \frac{\partial^2 \psi}{\partial x^2}$ do not depend on η and hence treated as constants in the integral.

2. In Eq. (17), the right side can be expanded as $\frac{1}{A}\int_{-\infty}^{+\infty} d\eta e^{\frac{im\eta^2}{2\epsilon\hbar}}\psi(t,x) +$ other terms. No terms on the right other than $\frac{1}{A}\int_{-\infty}^{+\infty} d\eta e^{\frac{im\eta^2}{2\epsilon\hbar}}\psi(t,x)$ has the form (constant) $\psi(x,t)$. On the right side of Eq. (17), you will find terms involving derivatives of ψ; you can even find a term in which ψ multiplies x, but you will not find any other "(constant) $\psi(t,x)$" term. Hence, $\frac{1}{A}\int_{-\infty}^{+\infty} d\eta e^{\frac{im\eta^2}{2\epsilon\hbar}}\psi(t,x)$, appearing on the right side of Eq. (17), must be equal to $\psi(t,x)$ appearing on the left side of the equation: $\psi(t,x) = \frac{1}{A}\int_{-\infty}^{+\infty} d\eta e^{\frac{im\eta^2}{2\epsilon\hbar}}\psi(t,x)$. Hence,

$\frac{1}{A}\int_{-\infty}^{+\infty} d\eta e^{\frac{im\eta^2}{2\epsilon\hbar}} = 1$, implying $A = \int_{-\infty}^{+\infty} d\eta e^{\frac{im\eta^2}{2\epsilon\hbar}} = \sqrt{\frac{2\pi i\hbar\epsilon}{m}}$ (In the last step, we used Eq. (10). Hence, $\frac{1}{A} = \sqrt{\frac{m}{2\pi i\hbar\epsilon}}$, same as what you got for a free particle! (Check the factor that multiplies the exponential in the expression of the free particle propagator in Example 1.)

3. $\frac{1}{A}\int_{-\infty}^{+\infty} d\eta e^{\frac{im\eta^2}{2\epsilon\hbar}}\eta = 0$, since the integrand is an odd function of η.

4. $\frac{1}{A}\int_{-\infty}^{+\infty} d\eta e^{\frac{im\eta^2}{2\epsilon\hbar}}\eta^2 = \frac{i\hbar\epsilon}{m}$, by using Eq. (11) and also using $\frac{1}{A} = \sqrt{\frac{m}{2\pi i\hbar\epsilon}}$ as obtained in Step 2.

We carry out the integrals on the right side of Eq. (17), mindful of the preceding steps 1 through 4, and keep terms up to $o(\epsilon)$, i.e., linear in ϵ. We obtain:

$$\psi(t,x) + \frac{i\hbar\epsilon}{2m}\frac{\partial^2 \psi(t,x)}{\partial x^2} + \frac{i}{\hbar}(\epsilon kx)\psi(t,x),$$ equating which to the left of Eq. (17), we get:

Chapter 6: Deriving Schrodinger's Equation using the Propagator

$$\psi(x,t) + \epsilon \frac{\partial \psi}{\partial t} = \psi(x,t) + \frac{i\hbar\epsilon}{2m}\frac{\partial^2 \psi(t,x)}{\partial x^2} + \frac{i}{\hbar}(\epsilon k x)\psi(t,x)$$

$$\Rightarrow \epsilon \frac{\partial \psi}{\partial t} = \frac{i\hbar\epsilon}{2m}\frac{\partial^2 \psi(t,x)}{\partial x^2} + \frac{i}{\hbar}(\epsilon k x)\psi(t,x)$$

Cancelling ϵ from both sides and multiplying both sides by $i\hbar$, we obtain:

$$i\hbar \frac{\partial \psi}{\partial t} = -\frac{\hbar^2}{2m}\frac{\partial^2 \psi}{\partial x^2} - kx\psi.$$

The equation we got is Schrodinger's equation for potential energy depending linearly on the space variable. The second term on the right side of the equation has linear potential energy $V(x) = -kx$. Hence, we can write Schrodinger's equation for this case as: $i\hbar \frac{\partial \psi}{\partial t} = -\frac{\hbar^2}{2m}\frac{\partial^2 \psi}{\partial x^2} + V(x)\psi$, which, as we will see, is the general form for Schrodinger's equation for any potential energy $V(x)$.

Case II: The Potential Energy $V(x)$ Is Quadratic in the Space Variable

Our derivation of Schrodinger's equation for the potential energy varying quadratically with the space variable will be similar to the case where the potential energy varied linearly with the space variable. Using the quadratic potential energy $V(x) = \frac{1}{2}kx^2$ in Eq. (13), we get:

$$K(t+\epsilon, x; t, y) = \frac{1}{A} e^{\frac{i}{\hbar}\left[\frac{m(x-y)^2}{2\epsilon} - \frac{1}{2}k\left(\frac{x+y}{2}\right)^2 \epsilon\right]}$$ [We replaced x in $V(x) = \frac{1}{2}kx^2$ by $\frac{x+y}{2}$. So, $V\left(\frac{x+y}{2}\right) = \frac{1}{2}k\left(\frac{x+y}{2}\right)^2$]

Using the above expression for the infinitesimal-time propagator in Eq. (2), viz., $\psi(t+\epsilon, x) = \int_{-\infty}^{+\infty} dy K(t+\epsilon, x; t, y)\psi(t, y)$, we get:

$$\psi(t+\epsilon, x) = \frac{1}{A}\int_{-\infty}^{+\infty} dy\, e^{\frac{i}{\hbar}\left[\frac{m(x-y)^2}{2\epsilon} - \frac{1}{2}k\left(\frac{x+y}{2}\right)^2 \epsilon\right]} \psi(t, y) \quad \ldots\ldots\ldots\text{Eq. (18)}.$$

Feynman's Path Integral explained with basic Calculus

Defining $\eta \equiv y - x$, we have $d\eta = dy$ (treating x as a constant). η varies from $-\infty$ to $+\infty$. We write the integral in Eq. (18) as follows:

$$\psi(t + \epsilon, x) = \frac{1}{A}\int_{-\infty}^{+\infty} d\eta \, e^{\frac{im\eta^2}{\hbar 2\epsilon}} e^{-\frac{i}{\hbar}\epsilon\frac{k}{2}\left(x+\frac{\eta}{2}\right)^2} \psi(t, x + \eta) \quad \ldots\ldots \text{Eq. (19)}.$$

1. We Taylor-expand the left side of Eq. (19) as:

$$\psi(t + \epsilon, x) = \psi(t, x) + \epsilon \frac{\partial \psi}{\partial t} + \cdots \quad \text{(By following Eq. (3))}$$

2. We also Taylor-expand $\psi(t, x + \eta)$ in the integrand on the right of Eq. (19) as:

$$\psi(t, x + \eta) = \psi(t, x) + \eta \frac{\partial \psi}{\partial x} + \frac{\eta^2}{2}\frac{\partial^2 \psi}{\partial x^2} + \cdots$$

(By following Eq. (4))

3. In the integrand on the right of Eq. (19), we Taylor-expand $e^{-\frac{i}{\hbar}\epsilon\frac{k}{2}\left(x+\frac{\eta}{2}\right)^2}$ by using Eq. (10) of Chapter 1, viz. $e^u = 1 + u + \cdots$, and by setting $u = -\frac{i}{\hbar}\epsilon\frac{k}{2}\left(x + \frac{\eta}{2}\right)^2$, as follows:

$$e^{-\frac{i}{\hbar}\epsilon\frac{k}{2}\left(x+\frac{\eta}{2}\right)^2} = 1 - \frac{i}{\hbar}\epsilon\frac{k}{2}\left(x + \frac{\eta}{2}\right)^2 + \cdots$$

(Keeping terms up to $o(\epsilon)$))

$$= 1 - \frac{i}{\hbar}\epsilon\frac{k}{2}\left(x^2 + \eta x + \frac{\eta^2}{4}\right) + \cdots$$

[Using $(a + b)^2 = a^2 + 2ab + b^2$].

Using the 1, 2 and 3 in Eq. (19), we get:

$$\psi(t, x) + \epsilon \frac{\partial \psi}{\partial t} + \cdots$$

$$= \frac{1}{A}\int_{-\infty}^{+\infty} d\eta \, e^{\frac{im\eta^2}{\hbar 2\epsilon}} \left[1 - \frac{i}{\hbar}\epsilon\frac{k}{2}\left(x^2 + \eta x + \frac{\eta^2}{4}\right) + \cdots\right]$$

$$\left[\psi(t, x) + \eta \frac{\partial \psi}{\partial x} + \frac{\eta^2}{2}\frac{\partial^2 \psi}{\partial x^2} + \cdots\right]$$

$$= \frac{1}{A}\int_{-\infty}^{+\infty} d\eta \, e^{\frac{im\eta^2}{\hbar 2\epsilon}} \left[\psi(t, x) + \eta \frac{\partial \psi}{\partial x} + \frac{\eta^2}{2}\frac{\partial^2 \psi}{\partial x^2} - \frac{i}{\hbar}\epsilon\frac{kx^2}{2}\psi(t, x) + \ldots\right]$$

$$\ldots\ldots\ldots\ldots\ldots\ldots\ldots\ldots \text{Eq. (20)}.$$

111

Chapter 6: Deriving Schrodinger's Equation using the Propagator

While evaluating various terms in the integral in Eq. (20), we need to be aware of the following:

1. $x, t, \psi(t,x), \dfrac{\partial \psi}{\partial x}, \dfrac{\partial^2 \psi}{\partial x^2}$ do not depend on η and hence treated as constants.

2. In Eq. (20), $\dfrac{1}{A}\int_{-\infty}^{+\infty} d\eta \, e^{\frac{im\eta^2}{\hbar \, 2\epsilon}} \psi(t,x)$, appearing on the right, must be equal to $\psi(t,x)$ appearing on the left (Just as in Case I: the "linear potential energy case".): $\psi(t,x) = \dfrac{1}{A}\int_{-\infty}^{+\infty} d\eta \, e^{\frac{im\eta^2}{\hbar \, 2\epsilon}} \psi(t,x)$.

   Hence, $\dfrac{1}{A}\int_{-\infty}^{+\infty} d\eta \, e^{\frac{im\eta^2}{\hbar \, 2\epsilon}} = 1$, implying $A = \int_{-\infty}^{+\infty} d\eta \, e^{\frac{im\eta^2}{\hbar \, 2\epsilon}} = \sqrt{\dfrac{2\pi i \hbar \epsilon}{m}}$ (In the last step, we used Eq. (10). Hence, $\dfrac{1}{A} = \sqrt{\dfrac{m}{2\pi i \hbar \epsilon}}$

   (Once again, same as what you got for a free particle and also for a particle having a potential energy linear in the space-variable:)

3. $\dfrac{1}{A}\int_{-\infty}^{+\infty} d\eta \, e^{\frac{im\eta^2}{\hbar \, 2\epsilon}} \eta = 0$, since the integrand is an odd function of η.

4. $\dfrac{1}{A}\int_{-\infty}^{+\infty} d\eta \, e^{\frac{im\eta^2}{\hbar \, 2\epsilon}} \eta^2 = \dfrac{i\hbar\epsilon}{m}$, using Eq. (11) and also using $\dfrac{1}{A}$ as calculated in Step 2.

Carrying out the integrals on the right of Eq. (20), mindful of the preceding steps 1, 2, 3 and 4, and keeping terms up to $o(\epsilon)$. We obtain:

$\psi(t,x) + \dfrac{i\hbar\epsilon}{2m}\dfrac{\partial^2 \psi(t,x)}{\partial x^2} - \dfrac{i}{\hbar}\epsilon\left(\dfrac{1}{2}kx^2\right)\psi(t,x)$, equating which to the left side of Eq. (20), we get:

$$\psi(x,t) + \epsilon\dfrac{\partial \psi}{\partial t} = \psi(x,t) + \dfrac{i\hbar\epsilon}{2m}\dfrac{\partial^2 \psi(t,x)}{\partial x^2} - \dfrac{i}{\hbar}\epsilon\left(\dfrac{1}{2}kx^2\right)\psi(t,x)$$

$$\Rightarrow \epsilon\dfrac{\partial \psi}{\partial t} = \dfrac{i\hbar\epsilon}{2m}\dfrac{\partial^2 \psi(t,x)}{\partial x^2} - \dfrac{i}{\hbar}\epsilon\left(\dfrac{1}{2}kx^2\right)\psi(t,x)$$

Cancelling ϵ from both sides and multiplying both sides by $i\hbar$, we obtain

$$i\hbar\dfrac{\partial \psi}{\partial t} = -\dfrac{\hbar^2}{2m}\dfrac{\partial^2 \psi}{\partial x^2} + \dfrac{1}{2}kx^2\psi$$

Feynman's Path Integral explained with basic Calculus

The above is Schrodinger's equation for quadratic potential energy. The second term on the right side of the above equation has potential energy $V(x) = \frac{1}{2}kx^2$. Hence, we can write Schrodinger equation for this case as:

$$i\hbar \frac{\partial \psi}{\partial t} = -\frac{\hbar^2}{2m}\frac{\partial^2 \psi}{\partial x^2} + V(x)\psi \text{ (once again.)}$$

Case-III: Deriving Schrodinger's Equation for a Particle with a General Potential Energy V(x)

In this final case, we will derive Schrodinger's equation for any general potential energy. This is the derivation Feynman referred to anecdotally in his Nobel lecture. Using Eq. (13) in Eq. (2), we get:

$$\psi(t+\epsilon, x) = \frac{1}{A}\int_{-\infty}^{+\infty} dy\, e^{\frac{i}{\hbar}\left[\frac{m(x-y)^2}{2\epsilon} - \epsilon V\left(\frac{x+y}{2}\right)\right]}\psi(t,y) \quad \ldots\ldots\ldots \text{Eq. (21)}.$$

Defining $\eta \equiv y - x$, we have $d\eta = dy$ (treating x as a constant). η, like y, varies from $-\infty$ to $+\infty$. Hence, in terms of η, the integral in Eq. (21) assumes the following form:

$$\psi(t+\epsilon, x) = \frac{1}{A}\int_{-\infty}^{+\infty} d\eta\, e^{\frac{im\eta^2}{\hbar\, 2\epsilon}} e^{-\frac{i}{\hbar}\epsilon V\left(x+\frac{\eta}{2}\right)}\psi(t, x+\eta) \quad \ldots\ldots\ldots \text{Eq. (22)}.$$

1. We Taylor-expand the left side as:

$$\psi(t+\epsilon, x) = \psi(x, t) + \epsilon \frac{\partial \psi}{\partial t} + \cdots \quad \text{[By following Eq. (3)]}$$

2. We also Taylor-expand $\psi(t, x+\eta)$ in the integrand on the right as

$$\psi(t, x \mid \eta) = \psi(t, x) + \eta \frac{\partial \psi}{\partial x} + \frac{\eta^2}{2}\frac{\partial^2 \psi}{\partial x^2} + \cdots$$

[By following Eq. (4)]

3. In the integrand on the right of Eq. (22), we Taylor-expand $e^{-\frac{i}{\hbar}\epsilon V\left(x+\frac{\eta}{2}\right)}$ by using Eq. (10) of Chapter 1, viz. $e^u = 1 + u + \cdots$, and by setting $u = -\frac{i}{\hbar}\epsilon V\left(x + \frac{\eta}{2}\right)$, as follows:

Chapter 6: Deriving Schrodinger's Equation using the Propagator

$$e^{-\frac{i}{\hbar}\epsilon V\left(x+\frac{\eta}{2}\right)} = 1 - \frac{i}{\hbar}\epsilon V\left(x+\frac{\eta}{2}\right) + \cdots$$

(Keeping only terms up to $O(\epsilon)$)

$$= 1 - \frac{i}{\hbar}\epsilon\left[V(x) + \frac{\eta}{2}\frac{dV}{dx} + \frac{1}{2}\left(\frac{\eta}{2}\right)^2\frac{d^2V}{dx^2} + \cdots\right] + \cdots$$

[In the last step, we Taylor-expanded $V\left(x+\frac{\eta}{2}\right)$ by following Eq. (11) of Chapter 1, $\frac{\eta}{2}$ playing the role of $(x-a)$].

Using 1, 2 and 3 in Eq. (22), we get:

$$\psi(t,x) + \epsilon\frac{\partial\psi}{\partial t} + \cdots$$

$$= \frac{1}{A}\int_{-\infty}^{+\infty} d\eta\, e^{\frac{im\eta^2}{\hbar\,2\epsilon}}\left[1 - \frac{i}{\hbar}\epsilon\left[V(x) + \frac{\eta}{2}\frac{dV}{dx} + \frac{1}{2}\left(\frac{\eta}{2}\right)^2\frac{d^2V}{dx^2} + \cdots\right]\right]$$

$$\left[\psi(t,x) + \eta\frac{\partial\psi}{\partial x} + \frac{\eta^2}{2}\frac{\partial^2\psi}{\partial x^2} + \cdots\right]$$

$$= \frac{1}{A}\int_{-\infty}^{+\infty} d\eta\, e^{\frac{im\eta^2}{\hbar\,2\epsilon}}\left[\psi(t,x) + \eta\frac{\partial\psi}{\partial x} + \frac{\eta^2}{2}\frac{\partial^2\psi}{\partial x^2} - \frac{i}{\hbar}\epsilon V(x)\psi(t,x) + \cdots\right]$$

.... Eq. (23).

While evaluating various terms in the above integral, we need to be aware of the following:

1. $V(x), \frac{dV}{dx}, \frac{d^2V}{dx^2}, \psi(x,t), \frac{\partial\psi}{\partial x}, \frac{\partial^2\psi}{\partial x^2}$ do not depend on η and hence treated as constants when carrying out integrals.

2. In Eq. (23), the right side can be expanded as $\frac{1}{A}\int_{-\infty}^{+\infty} d\eta\, e^{\frac{im\eta^2}{\hbar\,2\epsilon}}\psi(t,x) +$ other terms. No terms on the right other than $\frac{1}{A}\int_{-\infty}^{+\infty} d\eta\, e^{\frac{im\eta^2}{\hbar\,2\epsilon}}\psi(t,x)$ has the form "(constant) $\psi(t,x)$". You will find terms involving derivatives of ψ or terms in which ψ multiplies $V(x)$ and its derivatives, but you will not find another "(constant) $\psi(t,x)$" type of term. Hence, $\frac{1}{A}\int_{-\infty}^{+\infty} d\eta\, e^{\frac{im\eta^2}{\hbar\,2\epsilon}}\psi(t,x)$, appearing on the right of Eq. (23), must be equal to $\psi(t,x)$ appearing on the left: $\psi(t,x) =$

Feynman's Path Integral explained with basic Calculus

$\frac{1}{A}\int_{-\infty}^{+\infty} d\eta e^{\frac{im\eta^2}{\hbar 2\epsilon}} \psi(t,x)$. Hence, $\frac{1}{A}\int_{-\infty}^{+\infty} d\eta e^{\frac{im\eta^2}{\hbar 2\epsilon}} = 1$, implying $A = \int_{-\infty}^{+\infty} d\eta e^{\frac{im\eta^2}{\hbar 2\epsilon}} = \sqrt{\frac{2\pi i \hbar \epsilon}{m}}$ [We used Eq. (10)]. Hence, $\frac{1}{A} = \sqrt{\frac{m}{2\pi i \hbar \epsilon}}$ (Once again, same as what you got for a free particle and the potential energies you tried previously. Since the potential energy in this example is general, we have now conclusively established that $\frac{1}{A}$ is independent of the potential energy.)

3. $\frac{1}{A}\int_{-\infty}^{+\infty} d\eta e^{\frac{im\eta^2}{\hbar 2\epsilon}} \eta = 0$, since the integrand is an odd function of η.

4. $\frac{1}{A}\int_{-\infty}^{+\infty} d\eta e^{\frac{im\eta^2}{\hbar 2\epsilon}} \eta^2 = \frac{i\hbar\epsilon}{m}$, using Eq. (11) and also using $\frac{1}{A}$ as calculated in Step 2.

We carry out the integrals on the right of Eq. (23), mindful of the preceding steps 1 through 4, and keep terms up to $o(\epsilon)$ i.e., linear in ϵ. We obtain:

$\psi(x,t) + \frac{i\hbar\epsilon}{2m}\frac{\partial^2 \psi(x,t)}{\partial x^2} - \frac{i}{\hbar}\epsilon V(x)\psi(x,t)$, equating which to the left of Eq. (23), we get:

$\psi(x,t) + \epsilon\frac{\partial \psi}{\partial t} = \psi(x,t) + \frac{i\hbar\epsilon}{2m}\frac{\partial^2 \psi(x,t)}{\partial x^2} - \frac{i}{\hbar}\epsilon V(x)\psi(x,t)$

$\Rightarrow \epsilon\frac{\partial \psi}{\partial t} = \frac{i\hbar\epsilon}{2m}\frac{\partial^2 \psi(x,t)}{\partial x^2} - \frac{i}{\hbar}\epsilon V(x)\psi(x,t)$

Cancelling ϵ from both sides of the above equation and then multiplying both sides by $i\hbar$, we obtain:

$i\hbar\frac{\partial \psi}{\partial t} = -\frac{\hbar^2}{2m}\frac{\partial^2 \psi}{\partial x^2} + V(x)\psi$ .... Eq. (24).

The above is Schrodinger's equation for a general potential energy $V(x)$.

Solving Schrodinger's equation is not the focus of this book. Nonetheless, we will end the chapter with a few words about how to solve Eq. (24) for various potential energies. First of all, since $V(x)$ does not depend on t, you

Chapter 6: Deriving Schrodinger's Equation using the Propagator

can separate the space and the time parts in Eq. (24) by assuming the following form for ψ:

$$\psi(x,t) = e^{-i\frac{Et}{\hbar}} f(x) \quad \text{Eq. (25).}$$

$\frac{\partial \psi}{\partial t}$ can be calculated by differentiating ψ in the above w.r.t t, treating x as a constant. Hence,

$$\frac{\partial \psi}{\partial t} = \frac{-iE}{\hbar} e^{-i\frac{Et}{\hbar}} f(x) \text{ [We used } \frac{d}{du}(e^{au}) = ae^{au}, \text{ where } a = -i\frac{E}{\hbar}]$$

$$\quad \text{Eq. (26).}$$

Similarly, $\frac{\partial^2 \psi}{\partial x^2} = e^{-i\frac{Et}{\hbar}} \frac{d^2 f}{dx^2}$ ....Eq. (27).

Plugging Eq. (26) and Eq. (27) in Eq. (24), and after some algebra that includes cancelling $e^{-i\frac{Et}{\hbar}}$ from both sides, we obtain the following differential equation:

$$Ef = -\frac{\hbar^2}{2m}\frac{d^2 f}{dx^2} + V(x)f \quad \text{Eq. (28).}$$

Eq. (28) is called the "time-independent Schrodinger's equation". As it turns out to be the case, E in the above equation represents the particle's energy, which, in general, has multiple values, each value producing a solution f to Eq. (28). For example, we state without proof that if you choose the quadratic potential energy for $V(x)$, i.e., $V(x) = \frac{1}{2}kx^2$, you get the following quantized energy levels $E_n = \left(n + \frac{1}{2}\right)\hbar\omega$, where $\omega = \sqrt{\frac{k}{m}}$, and n assumes the integer values such as $n = 0, 1, 2, ...$ etc. For each energy level (i.e., for each value of n), you have a corresponding solution for f in Eq. (28). Note that the energies in this example are not only multiple, but also discrete.

On the other hand, for a free particle, you get a continuous range of E, which we show next, by setting $V(x) = 0$ in Eq. (28):

$$Ef = -\frac{\hbar^2}{2m}\frac{d^2 f}{dx^2} \quad \text{Eq. (29).}$$

Feynman's Path Integral explained with basic Calculus

Can you think of an f that solves the above equation? Let us try $f = e^{ikx}$, where k is a parameter. $\frac{d^2 f}{dx^2} = -k^2 e^{ikx} = -k^2 f$ (Since $e^{ikx} = f$). Plugging $\frac{d^2 f}{dx^2} = -k^2 f$ in Eq. (29), we get $E e^{ikx} = \frac{\hbar^2 k^2}{2m} e^{ikx}$, which implies $E = \frac{\hbar^2 k^2}{2m}$. Note that the energy E depends continuously on the parameter k: for any k, you get an E. Contrast this with the previously stated discrete energy-levels for a quadratic potential energy: $E_n = \left(n + \frac{1}{2}\right)\hbar\omega$. So, a free particle has a continuous range of energies, whereas a particle with a quadratic potential energy can only have discrete energy-values.

Finally, solving the three-dimensional version of Eq. (28) with the Coulomb potential energy for V, you get the discrete energies of the electron in a hydrogen atom.

A significant part of a standard quantum mechanics textbook is devoted to solving time-independent Schrodinger's equation for various physical systems.

Chapter 7 Classical Probabilities and Quantum Probability Amplitudes

This chapter assumes basic familiarity with Probability Theory, specifically, Conditional Probabilities.

Introduction

The relationship between the wavefunction and the propagator, viz., $\psi(B) = \int_{-\infty}^{+\infty} dx_c K(B,C) \psi(C)$ (Eq. (1)) in the previous chapter), is reminiscent of the probability theory. In classical probability theory, the conditional probability $P(B|C)$ is the probability that an event B happens given that the event C has already happened. If there are multiple events such as C_1, C_2, C_3 ......etc., there are multiple alternative ways in which B can happen, with $P(B|C_1)$, $P(B|C_2), P(B|C_3)$, ...etc. being corresponding conditional probabilities (See Figure 32).

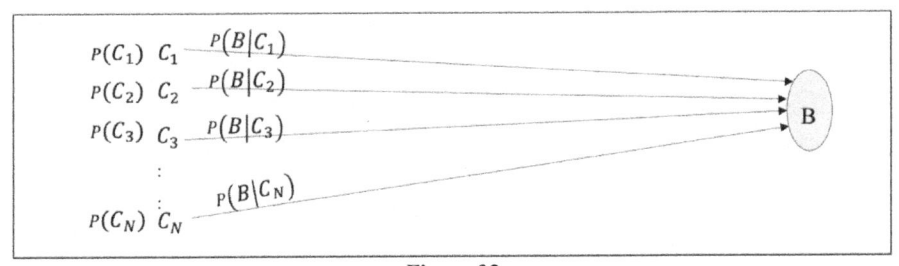

Figure 32

The probability of B can then be written as:
$P(B) = P(B|C_1)P(C_1) + P(B|C_2)P(C_2) + \cdots$, where $P(C_1), P(C_2), \ldots$ are probabilities of events $C_1, C_2, \ldots$ etc.

Feynman's Path Integral explained with basic Calculus

Writing compactly, $P(B) = \sum_{all\ C} P(B|C)P(C)$. Now, if x_c denotes a continuous variable associated with the event C, the sum in the previous expression can be replaced by the following integral:

$$P(B) = \int P(B|C)P(C)dx_c \quad \text{...............Eq. (1).}$$

We will give you concrete examples of how Eq. (1) is used to solve physical problems, but first notice the similarity between Eq. (1) and the key-relation from the previous chapter (Eq. (1), Chapter 6), viz., $\psi(B) = \int_{-\infty}^{+\infty} dx_c K(B,C) \psi(C)$: In the two formulae, parallels exists between $P(B)$ and $\psi(B)$ and between $P(C)$ and $\psi(C)$. Also, the propagator $K(B,C)$ is like $P(B|C)$. But the equivalences are purely formal. Still, it is titillating, since in one case you deal with probabilities and in another, probability amplitudes. In this chapter, you will see examples of Eq. (1) involving Gaussian integrals, similar in flavor to the type of math you saw in other chapters. The reasons for including this chapter in the book is:

1. to show you that the Gaussian integrals, essential for evaluating Path Integrals, show up elsewhere in science, and
2. a funny resemblance, although purely formal, exists between Eq. (1) and the relationship between the wave-function and the propagator.

RECAP: Formal Similarity Between Conditional Probability and Propagator

Probability Theory: $\quad P(B) = \int dx_c P(B|C)P(C)$

Quantum Mechanics: $\psi(B) = \int_{-\infty}^{+\infty} dx_c K(B,C) \psi(C)$

The propagator $K(B,C)$ in quantum mechanics is " formally like" $P(B|C)$.

Chapter 7: Classical Probabilities and Quantum Probability Amplitudes

Example 1: Digital PCR, Calculating Concentration from the Odds of Negatives

In this section, we will show you examples of Eq. (1). Our first example will be is from the field of biotechnology. Our discourse will be based on the work of Majumdar et. al (Poisson Plus Quantification for Digital PCR systems, Nature Scientific Reports, published on 29th August 2017). But to set the stage, let us consider the following problem. Imagine a very busy patch of a highway. The average number of car-accidents happening there can be of interest. To calculate the average, you can simply divide the total number of car-accidents taking place, say over a year, by the number of days in a year. Alternately, you can calculate the fraction (P_{neg}) of days which were accident-free over the span of a year and can find the average from P_{neg} by using the following simple formula that connects the two:

$$P_{neg} = e^{-average}.$$

Yes, it is possible to calculate the average from the fraction of negatives (which is also the odds of negatives) using the above formula! (Never mind why calculate average in this round-about way, for now. We will address that later.). We will show you how to derive $P_{neg} = e^{-average}$ shortly. But for now, let us demonstrate how the formula is useful for finding the average traffic accident. Setting the fraction of the days that were accident-free, i.e., P_{neg}, equal to $e^{-average}$, then solving for "average" from $P_{neg} = e^{-average}$, you get the "average number of traffic accidents per day" = $-\ln(P_{neg})$.

This powerful approach of calculating the average from the odds of negatives is employed in a cutting-edge concentration measurement technique used in biotechnology called Digital PCR. The goal of digital PCR is to detect diseases such as cancer, HIV or COVID. Certain types of DNA molecules (henceforth called target molecules), bearing the signature of disease, get released into the

Feynman's Path Integral explained with basic Calculus

patient's blood stream. The amounts of those molecules are quantified by measuring their concentration C, i.e., the number of target molecules per unit volume.

A processed blood sample containing target molecules is spread over thousands of micro-wells (See Figure 33). Target molecules go randomly into different wells, which means there can be 0, 1, 2, 3...basically any number of molecules in a given micro-well. Technology exists to make a micro-well fluoresce if it contains at least one target molecule in it. Non-fluorescing wells on the other hand contain zero target molecules. But the technology doesn't provide details as to exactly how many target molecules are present in a fluorescing well. All we know is that a fluorescing well contains at least one target molecule.

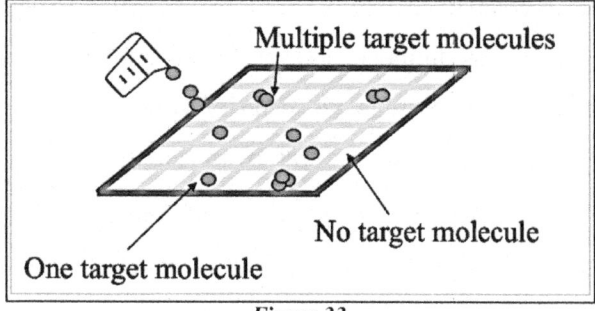

Figure 33

Now, this seems disappointing, since if we could count the exact number of molecules in each well, we could have added them up to get the total number of target molecules. Dividing that by the number of wells, we would then have obtained a*verage # of target molecules per well*, and then could have used the following formula to calculate the concentration C.

$$C = \frac{average\ \#\ of\ target\ molecules\ per\ well}{The\ volume\ V\ of\ a\ well} \quad \ldots\ldots\ldots\ldots\ldots\ldots\ldots \text{Eq. (2)}.$$

But no need to despair; the concentration can be calculated in an alternative way: by using $P_{neg} = e^{-average}$. Here is how. Although we cannot count the exact number of target molecules in each well, we can count how many wells didn't receive any target molecules using the existing technology (the wells with no molecules will simply not fluoresce). The fraction of the wells that did

Chapter 7: Classical Probabilities and Quantum Probability Amplitudes

not receive any molecules (the negatives) gives P_{neg}, which, when used in $P_{neg} = e^{-average}$ (the derivation of which we have not included yet), gives the $average \equiv average\ \#\ of\ target\ molecules\ per\ well$. Finally, the concentration C is obtained by dividing this average by the volume of a well.

The Derivation of: $P_{neg} = e^{-average}$

In this section, we will show how to connect the average with the probability of non-occurrence of an event (i.e., prove $P_{neg} = e^{-average}$), and then resume our main discussion in the next section.

Consider the tossing of a biased coin (meaning that the probability of the occurrence of head is different from the occurrence of tail.) N number of times. Let p be the probability that head occurs when the coin is tossed once. Hence $1 - p$ is the probability that a head does not occur. Multiplying $1 - p$ N times will give us the probability P_{neg} that no heads occur in N tosses.

Hence,

$$P_{neg} = (1-p)(1-p) \ldots\ldots N\ times = (1-p)^N \qquad \ldots\ldots\ldots\ \text{Eq. (3)}.$$

Using Binomial expansion (Eq. (7) of Chapter 1),

$$(1-p)^N = 1 - Np + \frac{N(N-1)}{2!}p^2 - \frac{N(N-1)(N-2)}{3!}p^3 + \cdots \quad \ldots\ldots\text{Eq. (4)}.$$

Now, let us take a look at, say, the term $\frac{N(N-1)}{2!}p^2 = \frac{Np(Np-p)}{2!}$. We assume p to be quite small, while N being sufficiently large such that $Np \equiv \lambda$ is of a "reasonable" magnitude. Then $Np - p \approx Np$ (since p is small). So, $\frac{Np(Np-p)}{2!} \approx \frac{(Np)(Np)}{2!} \equiv \frac{\lambda^2}{2!}$. Similarly, we will have $\frac{N(N-1)(N-2)}{3!}p^3 \approx \frac{(Np)^3}{3!} = \frac{\lambda^3}{3!}$, and so on.

Hence, from Eq. (4), we get $(1-p)^N \approx 1 - \lambda + \frac{\lambda^2}{2!} - \frac{\lambda^3}{3!} + \cdots = e^{-\lambda}$, using which in Eq. (3) we get $P_{neg} = e^{-\lambda}$, where $\lambda = Np$. Now, since p is the

Feynman's Path Integral explained with basic Calculus

probability that a head occurs and N the total number of tosses, $Np \equiv \lambda$ is the average number of times a head occurs in N coin-tosses. Hence, the equation $P_{neg} = e^{-\lambda}$ is actually $P_{neg} = e^{-average}$.

Although we derived $P_{neg} = e^{-average}$ using the example of coin-tossing, the formula is generic. To emphasize the point, if we replaced N number of coin-tosses in the above example by N target molecules (from our digital PCR problem), each having the probability p to enter into a well, we would have arrived at $P_{neg} = e^{-average}$ in exactly the same way. The average in that case would be the average number of target molecules per well.

You can calculate probabilities for any "generic" number of successes (not just the probability of zero successes or "negatives".), and obtain what is called a "Poisson" probability distribution (We will not get into the discussion of Poisson, since for our purpose, we need only a special case of Poisson, i.e., zero successes or occurrences. We brought up "Poisson", since the name appears in our next section. You do not need any understanding of the Poisson distribution (beyond what has already been discussed) to follow the rest.

Poisson Plus—A Digital PCR Algorithm

We have explained how $P_{neg} = e^{-average}$ can be used in digital PCR to calculate the average number of target molecules per well, and finally, the concentration C can be obtained using Eq. (2), reproduced here:

$$C = \frac{average\ \#\ of\ target\ molecules\ per\ well}{The\ volume\ V\ of\ a\ well}.$$

Problem!!! We can use the above formula to calculate concentration, when the volumes of all the wells are the same. In reality, the amount of liquid going into the microwells (Figure 33) may not be the same. This effectively translates into unequal microwell-volumes having an average of V_0 and a spread (i.e.,

Chapter 7: Classical Probabilities and Quantum Probability Amplitudes

standard deviation of σ), which are measured experimentally and treated as "known" parameters for the system.

The question then becomes what volume to use to calculate the concentration C.

Solution to the Problem

The solution to the problem is to modify $P_{neg} = e^{-average}$ in a way that the new formula accounts for the volume being variable, instead of being constant. Of course, the new formula should reduce to the old one when the volume is constant. Since the old formula is related to the Poisson distribution, and the new one is a modification, we refer to the new (yet to be derived) formula as Poisson Plus. [Once again, treat "Poisson" as just a label. "Plus" indicates modification.]

For a given well volume V, the average number of target molecules in a well is CV, C being the concentration, i.e., the average number of molecules per unit volume. Hence e^{-CV} is the conditional probability $P_{neg}|V$, i.e., the probability of negatives for a given volume V. So, $P_{neg}|V = e^{-CV}$, a straightforward application of $P_{neg} = e^{-average}$.

The volume V itself has a distribution $P(V)$, which we assume to be a Normal distribution given by $P(V) = \frac{1}{\sigma\sqrt{2\pi}} e^{-\frac{(V-V_0)^2}{2\sigma^2}}$, where V_0 and σ are the mean and the standard deviation of $P(V)$. Note that $P(V)$ is nothing but a normalized Gaussian you saw in Eq. (3) from Chapter 1.

Imagine for a moment that the variable V assumes only discrete values such as $V_1, V_2, V_3, \ldots \ldots$

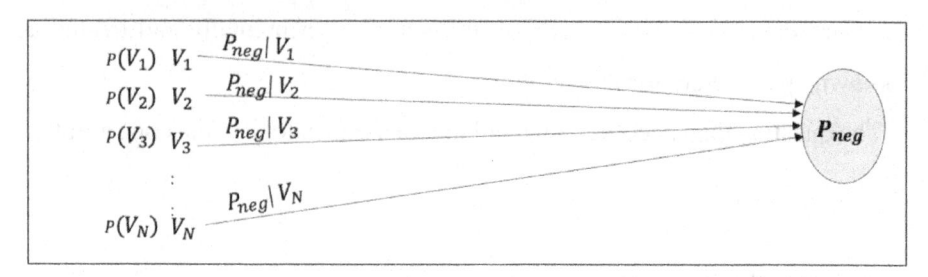

Feynman's Path Integral explained with basic Calculus

Figure 34

The probability of P_{neg} can then be written as (See Figure 34):

$$P_{neg} = (P_{neg}|V_1)P(V_1) + (P_{neg}|V_2)P(V_2) + \cdots + (P_{neg}|V_N)P(V_N)$$

But V being a continuous variable, we need to use the continuous version of the equation (along the lines of Eq. (1)), as follows.

$$P_{neg} = \int_{-\infty}^{+\infty}(P_{neg}|V)P(V)dV \quad \text{............................. Eq. (5).}$$

Using $P(V) = \frac{1}{\sigma\sqrt{2\pi}} e^{-\frac{(V-V_0)^2}{2\sigma^2}}$ and $P_{neg}|V = e^{-CV}$ in the above, we get

$$P_{neg} = \frac{1}{\sigma\sqrt{2\pi}} \int_{-\infty}^{+\infty} e^{-CV} e^{-\frac{(V-V_0)^2}{2\sigma^2}} dV$$

$$= \frac{1}{\sigma\sqrt{2\pi}} \int_{-\infty}^{+\infty} e^{-\left[CV + \frac{(V-V_0)^2}{2\sigma^2}\right]} dV \quad \text{........................... Eq. (6).}$$

Note that in the integral in Eq. (6), the lower and the upper limits of the volume variable are $-\infty$ and $+\infty$, which is clearly unrealistic since volumes can never be negative. But it's a good approximation if the mean volume V_0 is much larger than the spread or the standard deviation σ, which is an entirely realistic assumption to make. Then, as shown in Figure 35, the Gaussian function becomes practically zero for all values of the volume far away from V_0,

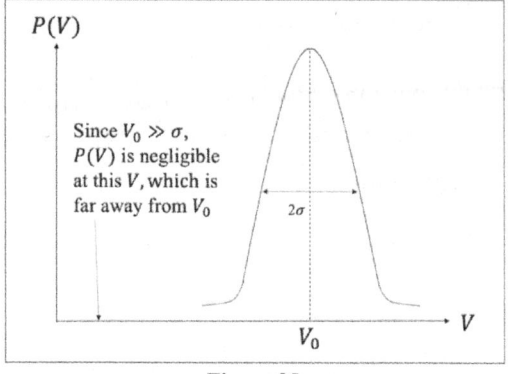

Since $V_0 \gg \sigma$, $P(V)$ is negligible at this V, which is far away from V_0

Figure 35

Chapter 7: Classical Probabilities and Quantum Probability Amplitudes

including the negative values. So, we do not incur error by taking the limits of integration from minus to plus infinity. If V_0 is not significantly larger than σ, more realistic models for the volume distribution $P(V)$ (such as the "truncated Gaussian") which are zero for negative values of the volume, can be used.

By expanding the square in the argument of the exponential in the integrand of Eq. (6), you can write the integral (after some algebra) in the form: (constant) $\int_{-\infty}^{+\infty} e^{-AV^2+BV} dV$, and evaluate the integral using Eq. (6) of Chapter 1, viz., $\int e^{-AV^2+BV} dV = \sqrt{\frac{\pi}{A}} e^{\frac{B^2}{4A}}$. Finally, after taking these steps, you obtain from Eq. (6) of the current chapter:

$$P_{neg} = e^{(\frac{1}{2}\sigma^2 C^2 - CV_0)} \quad \text{... Eq. (7).}$$

(You will be asked to derive Eq. (7) in Exercise 1). Taking logarithm on both sides of Eq. (7), we get:

$$ln(P_{neg}) = \frac{1}{2}\sigma^2 C^2 - CV_0.$$

In the above equation, P_{neg}, σ and V_0 are known quantities. Hence, everything in the above equation is known except for C. Solving for C using the quadratic equation formula, we obtain the following expression for the concentration.

$$C = \frac{V_0 - V_0\sqrt{1+2\frac{\sigma^2}{V_0^2}ln(P_{neg})}}{\sigma^2} \quad \text{.................................... Eq. (8).}$$

The above is the Poisson Plus formula used in the digital PCR technology. [We ignored the root with the positive sign in front of the radical; the reason for doing that will be explained shortly.]

When σ in Eq. (8) is small, meaning the volumes of all the micro-wells are approximately equal to V_0, one can Taylor-expand $\sqrt{1 + 2\frac{\sigma^2}{V_0^2}ln(P_{neg})} \approx 1 +$

Feynman's Path Integral explained with basic Calculus

$\frac{\sigma^2}{V_0^2} ln(P_{neg})$ by using the formula $(1 + x)^{\frac{1}{2}} \approx 1 + \frac{1}{2}x$ (Eq. (8) from Chapter 1). Hence, we obtain from Eq. (8) of the current chapter,

$$C \approx \frac{1}{\sigma^2}\left(V_0 - V_0\left(1 + \frac{\sigma^2}{V_0^2} ln(P_{neg})\right)\right) = \frac{-ln(P_{neg})}{V_0} \quad \ldots\ldots\ldots\ldots \text{Eq. (9).}$$

C, as given above, is independent of the standard deviation σ and is precisely what one gets by solving the formula $P_{neg} = e^{-CV_0}$ for a "fixed volume $V = V_0$" case, and thus reproduces the previous result.

[Note: If in the formula for C in Eq. (8), we used a positive sign in front of the square root, C wouldn't have reduced to Eq. (9) in the limiting case of vanishing σ. Anticipating that, the positive sign had been discarded in Eq. (8).]

Exercise 1. Carry out the integral in Eq. (6) and derive Eq. (7).

RECAP: <u>Poisson Plus Algorithm (from biotechnology)</u>

- $P_{neg} = \int_{-\infty}^{+\infty}(P_{neg}|V)P(V)dV$, where

  - $P_{neg}|V = e^{-CV}$ [Follows from $P_{neg} = e^{-average}$]

  - $C = \dfrac{Average\ \#\ of\ target\ molecules\ per\ well}{The\ volume\ V\ of\ a\ well}$

  - $P(V) = \dfrac{1}{\sigma\sqrt{2\pi}} e^{-\frac{(V-V_0)^2}{2\sigma^2}}$ [Normalized Gaussian with Average V_0 and Standard Deviation σ]

- The mean volume V_0 is much larger than the standard deviation σ.

Chapter 7: Classical Probabilities and Quantum Probability Amplitudes

Example 2: Random Walk

As a second example of how $P(B) = \int P(B|C)P(C)dx_c$, as given by Eq. (1), works, we consider the following. Imagine you are driving a car at a constant speed. The distance the car covers at time t is proportional to t, right? Now, imagine a drunkard in a very narrow, straight alley. He starts walking from a pole located somewhere in the alley (as shown in Figure 36) and takes steps randomly to the right and to the left.

Pole: The drunkard starts here.

The drunkard is here after a time t.

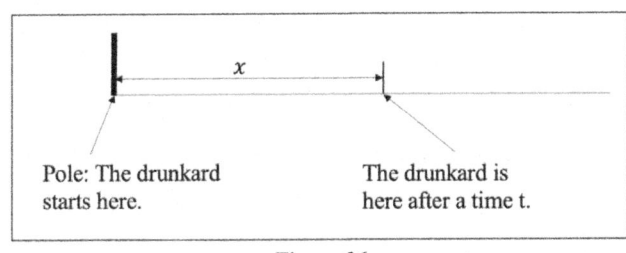

Figure 36

Since he is too drunk to make a choice, his odds of going either to the right or left is the same.

Can you tell exactly where he will be at time t after he starts walking? Of course, not. However, we can construct a probability distribution function $P(x,t)$ for the drunkard's being at the position x, at time t. It appeals to the intuition that $P(x,t)$ is symmetrical about $x = 0$ (the drunkard's starting point), since the drunkard is as likely to go to the left as to the right (See Appendix 6 more details). Hence, if we approximate $P(x,t)$ by a Gaussian or a Normal distribution, its center or mean will be located at $x = 0$. You can rigorously prove that $P(x,t)$ is Gaussian when the total number of steps N the drunkard takes is large and the length of each step l is small; (We will not get into proving that). You can also show that the standard deviation of such a Gaussian is proportional to the square root of time: $\sqrt{t}$, i.e., equal to $D\sqrt{t}$, where D is the proportionality constant. (See Appendix 6 more details). It means that the distribution will become more spread with time.

Feynman's Path Integral explained with basic Calculus

Now, the zero mean Gaussian with a standard deviation σ is given by $P(x,t) = \frac{1}{\sigma\sqrt{2\pi}} e^{-\frac{x^2}{2\sigma^2}}$. Substituting $\sigma = D\sqrt{t}$, we get:

$$P(x,t) = \frac{1}{D\sqrt{t}\sqrt{2\pi}} e^{-\frac{x^2}{2D^2 t}} \quad\ldots\ldots\ldots\ldots\ldots\text{ Eq. (10).}$$

Eq. (10) assumes that the drunkard started at $x = 0$ at time $t = 0$. But suppose he had started walking from $x = x_a$ at a time $t = t_a$ instead. Then the probability that he reaches x_b at time $t = t_b$ can be obtained by replacing x and t on the right of Eq. (10) by $x_b - x_a$ and by $t_b - t_a$ respectively. Hence, if we denote such a probability by $P(t_b, x_b; t_a, x_a)$, in which, the time and space co-ordinates of the final position are written before those of the initial position in the notation, we get:

$$P(t_b, x_b; t_a, x_a) = \frac{1}{D\sqrt{t_b - t_a}\sqrt{2\pi}} e^{-\frac{1}{2D^2}\frac{(x_b - x_a)^2}{(t_b - t_a)}} \quad\ldots\ldots\ldots\ldots\text{ Eq. (11).}$$

Compare Eq. (11) with the quantum propagator for a free particle in Chapter 4, Eq. (5). Notice their formal similarity except in the occurrence of i.

Next, suppose that the drunkard goes from the point A, indicated by (t_a, x_a) in the space-time diagram to $B(t_b, x_b)$, via $C(t_c, x_c)$ (Figure 37). C is an arbitrary intermediate point, its spatial co-ordinate x_c can assume any value. So, we are dealing with three probabilities here: the probability to go from A to C (denoted by $P(C)$); the probability to go from C to B (denoted by $P(B|C)$); and the probability to go from A to B (denoted by $P(B)$). The probabilities, mentioned above, should be related to each other as

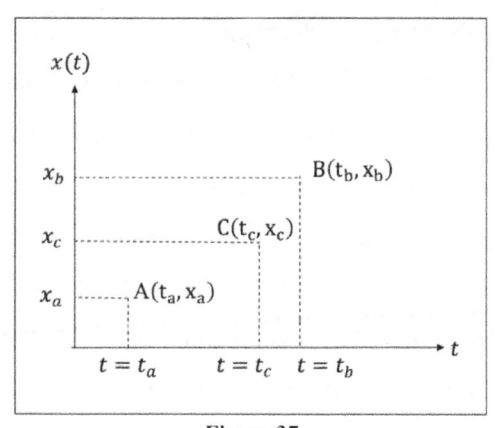

Figure 37

129

Chapter 7: Classical Probabilities and Quantum Probability Amplitudes

in Eq. (1), which is the general rule of combining probabilities: $P(B) = \int P(B|C)P(C)dx_c$.

We will write $P(B)$, $P(C)$ and $P(B|C)$ in a more notationally detailed way as follows:

$P(B) \equiv P(t_b, x_b; t_a, x_a)$, where in the notation, the co-ordinates (t_b, x_b) of the destination point B comes before the co-ordinates (t_a, x_a) of the initial point A. (Remember that in the notation $P(B)$, it is implied that the particle starts at A). From Eq. (11), we have:

$$P(t_b, x_b; t_a, x_a) = \frac{1}{D\sqrt{t_b - t_a}\sqrt{2\pi}} e^{-\frac{1}{2D^2}\frac{(x_b - x_a)^2}{(t_b - t_a)}}.$$

Similarly, $P(C) \equiv P(t_c, x_c; t_a, x_a)$, where $C(t_c, x_c)$ and $A(t_a, x_a)$ are the final and the initial points respectively. Hence, along the lines of Eq. (11) with the correct initial and final co-ordinates, we have:

$$P(t_c, x_c; t_a, x_a) = \frac{1}{D\sqrt{t_c - t_a}\sqrt{2\pi}} e^{-\frac{1}{2D^2}\frac{(x_c - x_a)^2}{(t_c - t_a)}} \quad \ldots\ldots\ldots\ldots\ldots \text{Eq. (12)}.$$

Finally, the conditional probability $P(B|C) \equiv P(t_b, x_b; t_c, x_c)$ denotes the particle's probability of reaching B from C, and hence the co-ordinates of B show up before those of C in the notation. Reasoning along the lines of Eq. (11) and Eq. (12), we get:

$$P(t_b, x_b; t_c, x_c) = \frac{1}{D\sqrt{t_c - t_b}\sqrt{2\pi}} e^{-\frac{1}{2D^2}\frac{(x_c - x_b)^2}{(t_c - t_b)}} \quad \ldots\ldots\ldots\ldots\ldots \text{Eq. (13)}.$$

Our interest is to explicitly check if Eq. (1), i.e., $P(B) = \int P(B|C)P(C)dx_c$ holds, which, per the recently introduced notations, is equivalent to checking if the following is true:

$$P(t_b, x_b; t_a, x_a) = \int_{-\infty}^{+\infty} dx_c P(t_b, x_b; t_c, x_c) P(t_c, x_c; t_a, x_a)$$

$$\ldots\ldots\ldots\ldots\ldots\ldots\ldots\ldots \text{Eq. (14)},$$

where the probabilities $P(t_b, x_b; t_a, x_a)$, $P(t_c, x_c; t_a, x_a)$, and $P(t_b, x_b; t_c, x_c)$ are given by Eq. (11), Eq. (12) and Eq. (13) respectively. On

Feynman's Path Integral explained with basic Calculus

the right of Eq. (14), you need to carry out the integration
$\int_{-\infty}^{+\infty} dx_c e^{-\frac{1}{2D^2}\frac{(x_b-x_c)^2}{(t_b-t_c)} - \frac{1}{2D^2}\frac{(x_c-x_a)^2}{(t_c-t_a)}}$, for which you can use the following formula you encountered in chapter 1 (Eq. (16)):

$$\int_{-\infty}^{+\infty} dx e^{[-k_1(x-a)^2 - k_2(x-b)^2]} = \sqrt{\frac{\pi}{k_1+k_2}} e^{\frac{-k_1 k_2}{k_1+k_2}(a-b)^2}.$$

After doing the integration in Eq. (14) of the current chapter, and carrying out subsequent algebraic simplifications, you should get what you have on the left side of Eq. (14).

Exercise 1. Prove Eq. (14) following the steps outlined above.

The random walking of a drunkard discussed in this chapter is used to model the random motion of pollen in water. The zigzag motion of the pollen, resulting from being bombarded by the water molecules that it is surrounded by, is called the Brownian motion. A pollen's movement, happening on the water surface, i.e., in two dimensions, is an example of a two-dimensional Brownian motion. Eq. (11) gives the probability distribution of a particle executing a one-dimensional Brownian motion.

Appendix 1

Functionals

We know a function is an input-output relationship. A function takes a number as an input and spits out another as an output. For example, for the input $x = 3$, the function $f(x) = x^2$ assumes the value $f(3) = 3^2 = 9$.

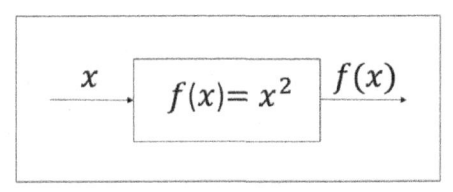

A "functional" takes an entire function as the input and spits out a single number as the output. For example, an integral, such as $\int_{t_a}^{t_b} x^2(t)dt$ is a functional, since for the function $x(t)$ defined over an interval of t, the integral produces a single number. Symbolically, we can write a functional as $F[x(t)]$. For our problem, $F[x(t)] = \int_{t_a}^{t_b} x^2(t)dt$.

Let us evaluate $F[x(t)] = \int_{t_a}^{t_b} x^2(t)dt$ for $x(t) = t$, and $t_a = 0$, $t_b = 3$.

$$F[x(t)] = \int_{t_a}^{t_b} x^2(t)dt$$

$$= \int_0^3 t^2 dt \quad [\text{Since } x(t) = t, x^2(t) = t^2]$$

$$= \left.\frac{t^3}{3}\right|_{t=0}^{t=3} = 9$$

Next, let us evaluate another functional $F[x(t)] = \int_0^3 \left(\frac{dx}{dt}\right)^2 dt$ for the same function $x(t)$, i.e., $x(t) = t$.

Feynman's Path Integral explained with basic Calculus

Since, $x(t) = t$, $\frac{dx}{dt} = 1$. Hence, $F[x(t)] = \int_0^3 \left(\frac{dx}{dt}\right)^2 dt = \int_0^3 1 dt = 3 - 0 = 3$. So, $F[x(t)] = 3$ in this case.

The arc-length of an arbitrary curve is also a functional since the arc length involves an integral and depends on the curve (function) in question.

Appendix 2

Potential Energies

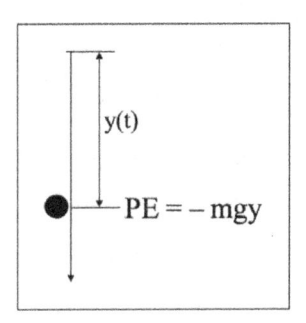

We are familiar with Newton's law: $F = ma$, where F is the force on an object of mass m, and a is the acceleration. For example, gravitational force pulls a mass m down with an acceleration g. Hence the gravitational force is written as: $F = mg$ [We chose our co-ordinate system such that the downward direction is positive y]. Is there a function, the negative derivative of which with respect to the position variable is the force? Let us try $V = -mgy$. $-\frac{dV}{dy} = mg$, which is equal to the gravitational force on an object of mass m! $V = -mgy$ is called the gravitational potential energy. The relationship between the force F and the potential energy V is $F = -\frac{dV}{dy}$: Force is the negative of the derivative of the potential energy with respect to the space variable.

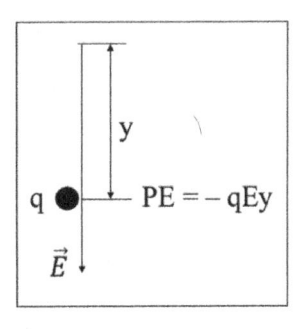

Next, in the above problem, replace the gravitational force by an electrostatic force. A charge q in an electric field E acting downward (the positive y direction) will experience a force $F = qE$. If q is positive, the force on the charge will be in the same direction as the electric field E, and if q is negative, the force on the charge will be oppositely directed to the electric field. We ask the same question as before: Is there a function, the negative derivative of which with respect to the position

Feynman's Path Integral explained with basic Calculus

variable, is the force? Let us try $V = -qEy$. $-\frac{dV}{dy} = qE$, which is equal to the electric force on an object of charge q! Hence $V = -qEy$ acts as the electric potential energy.

Next, consider a spring mass system. The force exerted on the mass by the spring is given by: $F = -kx$ (Hooke's law). The question is: what V satisfies $F = -\frac{dV}{dx} = -kx$, i.e., $\frac{dV}{dx} = kx$?

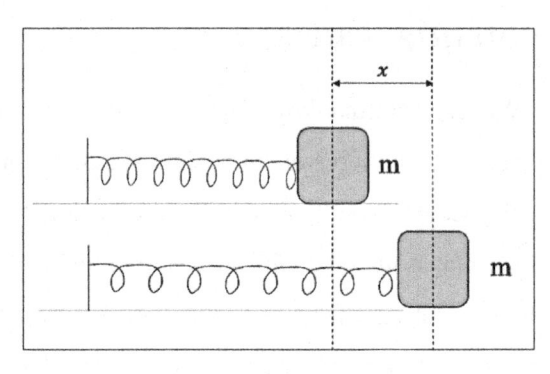

Instead of guessing, why not integrate $\frac{dV}{dx} = kx$? Integrating, you get $V = \frac{1}{2}kx^2$, the potential energy for the spring mass system.

What is the use of the potential energy? Well, as you saw in the book, the potential energy plays a crucial role in Feynman's Path Integral formalism. There are other important applications of potential energy, such as in the energy conservation: when an apple falls from a tree, the sum of its potential and kinetic energies remains constant at every instant during its fall (ignoring the air friction). In our book, however, the usefulness of the potential energy lies in its presence in a concept called "action", central to Path Integrals.

Appendix 3

Heisenberg's Uncertainty Principle: an Application of the Multiplicative Law of Propagators (Advanced Topic)

In this appendix, we will use Eq. (6) from Chapter 5, viz., $K(B,A) = \int_{-\infty}^{+\infty} dx_c K(B,C)K(C,A)$, to explain Heisenberg's uncertainty principle, which is of fundamental importance in quantum mechanics. We will follow closely the derivation provided in *Quantum Mechanics and Path Integrals* by Richard Feynman and Albert Hibbs. We will not derive all the mathematical expressions needed for the explanation. Refer to section 3.2 of the book mentioned above for a detailed derivation.

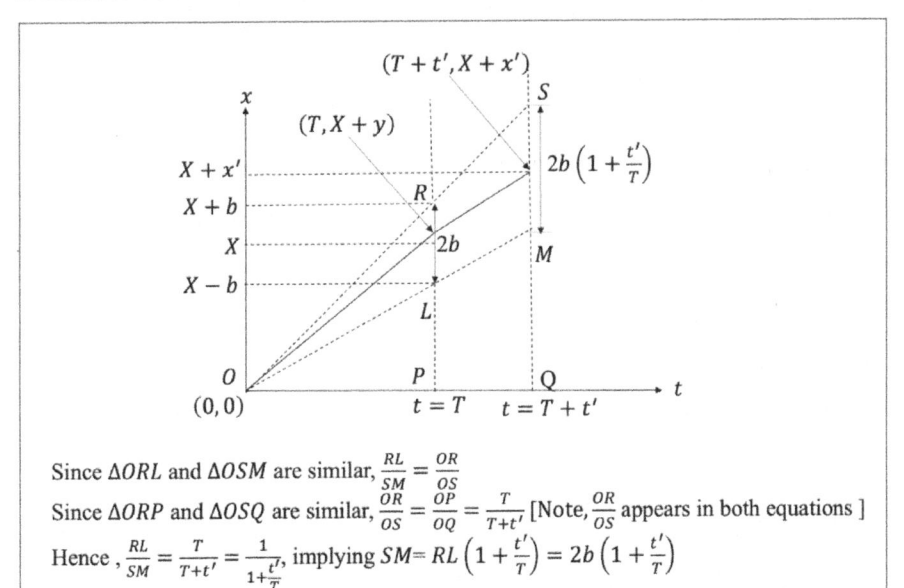

Since $\triangle ORL$ and $\triangle OSM$ are similar, $\frac{RL}{SM} = \frac{OR}{OS}$

Since $\triangle ORP$ and $\triangle OSQ$ are similar, $\frac{OR}{OS} = \frac{OP}{OQ} = \frac{T}{T+t'}$ [Note, $\frac{OR}{OS}$ appears in both equations]

Hence, $\frac{RL}{SM} = \frac{T}{T+t'} = \frac{1}{1+\frac{t'}{T}}$, implying $SM = RL\left(1+\frac{t'}{T}\right) = 2b\left(1+\frac{t'}{T}\right)$

Figure 38

Feynman's Path Integral explained with basic Calculus

Let us say a free particle starts at $x = 0$ at time $t = 0$ in the space-time diagram (Figure 38). Since the particle is free, if we knew the particle's speed with certainty, we could have exactly predicted its location at a later time $t = T$ (denoted by P on the time axis.). For example, if the particle's speed at $t = 0$ was $\frac{X}{T}$, its distance from the origin after a time T would be $(speed)(time) = \frac{X}{T} \cdot T = X$. But let us assume that for whatever reasons the particle's speed is not known with absolute certainty. As a result, all we know for sure is that the particle will be within $\pm b$ of X after a time T. The particle's trajectory, not exactly known, will lie between the lines OR and OL in the space-time diagram. RL, representing the particle's uncertainty in position at time T, is $2b$, as indicated in the figure. (X is the midpoint of RL). The question: Will the uncertainty magnify with time? (We haven't discussed anything "quantum" yet.)

The answer is yes. If the particle is at L at $t = T$, it will continue to move along OL and reach, say, the point M at a later time $t = T + t'$, denoted by Q on the time-axis. Whereas, if the particle is at R at $t = T$, it will continue to move along OR, and reach the point S at $t = T + t'$. The particle's actual trajectory is somewhere in between OS and OM. Hence, the uncertainty in the particle's location at $t = T + t'$ is SM. As you can see from the figure, SM is greater than RL, clearly demonstrating a larger uncertainty at a later time. We obtain an expression of SM in terms of RL, which is $2b$, in the following:

$\triangle ORL$ and $\triangle OSM$ are similar. Hence:

$$\frac{OR}{OS} = \frac{RL}{SM} \quad \text{..} \text{Eq. (1).}$$

Since $\triangle ORP$ and $\triangle OSQ$ are similar,

$$\frac{OR}{OS} = \frac{OP}{OQ} = \frac{T}{T+t'} \quad \text{................................} \text{Eq. (2).}$$

By exploiting the fact that $\frac{OR}{OS}$ is present in both Eq. (1) and Eq. (2), we obtain:

Appendix 3

$$\frac{RL}{SM} = \frac{T}{T+t'}$$

$$= \frac{1}{1+\frac{t'}{T}} \quad \text{.. Eq. (3).}$$

Eq. (3) implies $SM = RL\left(1 + \frac{t'}{T}\right)$

$$= 2b\left(1 + \frac{t'}{T}\right) \quad [\text{Using RL}=2b] \text{ .... Eq. (4).}$$

Eq. (4) gives the classical uncertainty in the particle's location. There is nothing quantum about this result; it simply demonstrates that if there is an uncertainty in pinning down a particle's location (for whatever reason) at any time, the uncertainty gets magnified with time, as given by Eq. (4) in the above.

But, our particle is subjected to the laws of quantum mechanics, where we speak in the language of probability amplitudes. If we can find the particle's probability amplitude to go from $(t = 0, x = 0)$ in the space-time graph to $(t = T + t', x = X + x')$, we can square this probability amplitude and obtain the corresponding probability distribution, the spread of the probability distribution curve giving the quantum uncertainty in the particle's position at the time $t = T + t'$.

The problem, described above, has two parts. (See Figure 38) First, you need to find the probability amplitude of the particle's going from $(t = 0, x = 0)$ to $(t = T, x = X + y)$, where y varies between $-b$ to $+b$, forcing x to stay within $X - b$ to $X + b$ at the time $t = T$.

In the second part of the problem, we find the probability amplitude for the particle's going from $x = X + y$ at time T to an arbitrary location $X + x'$ at time $T + t'$.

In the first part, the free particle propagator from $(t = 0, x = 0)$ to $(t = T, x = X + y)$ is obtained by setting $t_a = 0, x_a = 0$, and $t_b = T, x_b = X + y$ in Eq. (5) of Chapter 4. We get:

Feynman's Path Integral explained with basic Calculus

$$K(T, X+y; 0,0) = \sqrt{\frac{m}{2\pi i \hbar T}} e^{\frac{im(X+y)^2}{2\hbar T}} \quad \text{............................ Eq. (5)}$$

In the second part of the problem, the probability amplitude for the particle to go from $(t = T, x = X + y)$ to $(t = T + t', x = X + x')$ is obtained by setting $x_a = X + y, t_a = T$, and $x_b = X + x', t_b = T + t'$ in Eq. (5) of Chapter 4. We get:

$$K(T + t', X + x'; T, X + y) = \sqrt{\frac{m}{2\pi i \hbar t'}} e^{\frac{im(x'-y)^2}{2\hbar t'}} \quad \text{.............. Eq. (6).}$$

Next we will use the key formula of this appendix (mentioned at the beginning of the appendix), viz., $K(B, A) = \int_{-\infty}^{+\infty} dx_c K(B, C) K(C, A)$, where the initial point is given by $A \equiv (0, 0)$, the intermediate point is given by $C \equiv (T, X + y)$ and the final point is given by $B \equiv (T + t', X + x')$. Hence, $K(C, A)$ and $K(B, C)$ are given by Eq. (5) and Eq. (6) respectively. $K(B, A) \equiv K(T + t', X + x'; 0, 0)$ is the probability amplitude to go from $x = 0$ at $t = 0$ to $x = X + x'$ at $t = T + t'$. Hence, by using the above-mentioned key formula, we get:

$$K(T + t', X + x'; 0,0)$$
$$= \int_{-b}^{+b} K(T + t', X + x'; T, X + y) K(T, X + y; 0,0) \, dy$$

$$\text{... Eq. (7).}$$

where, the K's in the above integral are given by Eq. (5) and Eq. (6). Note that y varies between $-b$ and $+b$. Since we know with certainty that the particle is within $X \pm b$ at time T, the variable y associated with time T must vary between $-b$ and $+b$. Eq. (7) is hard to evaluate due to the limits of integration being finite (Sometimes

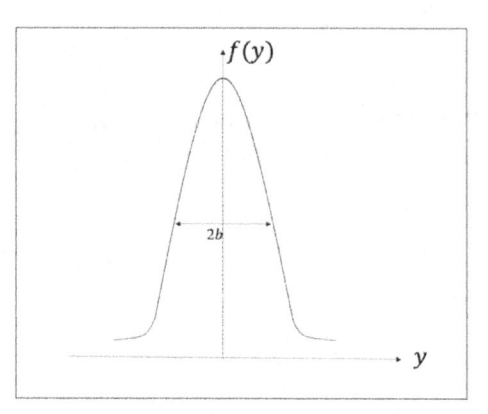

Appendix 3

infinities make things easier). We remedy that by multiplying the integrand of Eq. (7) by the Gaussian $G(y) = e^{-\frac{y^2}{2b^2}}$, and letting the limits of integration go from $-\infty$ to $+\infty$. $e^{-\frac{y^2}{2b^2}}$ has a width of $2b$, since it falls off quickly for absolute values of y greater than b (See the figure). Hence, multiplying the integrand of Eq. (7) by $e^{-\frac{y^2}{2b^2}}$ is, in some approximate sense, like having the limits of integration from $-b$ to $+b$.

After carrying out the integral in Eq. (7) with the suggested modification, and obtaining $K(T + t', X + x'; 0, 0)$, you can take the absolute square of $K(T + t', X + x'; 0, 0)$, which will give you the probability $P(T + t', X + x')$ of finding the particle at $X + x'$ at the time $T + t'$. We state the result as follows:

$$P(X + x', T + t') = \frac{m}{2\pi\hbar} \frac{b}{T\Delta x} e^{-\frac{\left(x' - \frac{X}{T}t'\right)^2}{(\Delta x)^2}} \quad \text{Eq. (8),}$$

where $(\Delta x)^2 = b^2 \left(1 + \frac{t'}{T}\right)^2 + \frac{\hbar^2 t'^2}{m^2 b^2}$ .... Eq. (9).

Now, Eq. (8) describes a Gaussian function centered at $x' = \frac{X}{T}t'$ and having a width whose square is given by Eq. (9). As you can see from Eq. (8) and Eq. (9), both the center and the width of the Gaussian change with the time t'. From Eq. (9), it is clear that the width or the spread increases with t'.

The spread-squared $(\Delta x)^2$ has two terms: the first one does not involve the Planck's constant $\hbar$, whereas the second one does. We will analyze the two terms separately, i.e., by setting the other term to zero when analyzing one. Setting the second term in Eq. (9) to zero, you get $\Delta x = \pm b \left(1 + \frac{t'}{T}\right)$, amounting to a net "uncertainty" of $2b \left(1 + \frac{t'}{T}\right)$ at $t = T + t'$, the uncertainty being exactly what we got in Eq. (4). So, the previously discussed non-quantum uncertainty is also present in the quantum formula.

Feynman's Path Integral explained with basic Calculus

The second term in Eq. (9), however, is purely quantum mechanical. Note that when $b \to 0$ how the first term becomes vanishing small and the second term dominates. In that limit, $\Delta x = \frac{\hbar t'}{mb}$ is the quantum uncertainty, dividing which by t' you get the corresponding uncertainty in the velocity as: $\Delta v = \frac{\Delta x}{t'} = \frac{\hbar}{mb}$. Hence, the uncertainty in the momentum is: $\Delta p = m\Delta v = \frac{\hbar}{b}$, multiplying which by the uncertainty in the position, i.e., $\Delta x = 2b$, we get $\Delta p \Delta x = 2\hbar \sim o(\hbar)$ (Heisenberg's uncertainty principle.)

Appendix 4

The Wide-Ranging Applicability of Free Particle Propagators (Advanced Topic)

We mentioned at the end of the section called "The Connection between Classical and Quantum" in Chapter 4 that free particle propagators are of importance even when a particle is not "free" but possesses a potential energy. In this appendix, we discuss that, although we will not go into the details and simply present an idea of the topic.

In Eq. (6) of Chapter 5, we stated the following expression involving propagators:

$$K(B,A) = \int_{-\infty}^{+\infty} dx_c K(B,C) K(C,A) \quad \text{...........................Eq. (1)}$$

Eq. (1) is valid whether or not a particle is free. But when a particle has a potential energy, you cannot use free particle propagators for the K's in Eq. (1); you need to use K's appropriate for the particle's potential energy. However, as it turns out, you can still use free particle propagators for calculating the probability amplitude $K(B,A)$, by means of a formalism derived from Eq. (1) using "Perturbation theory" (A topic we are not covering in this book.) The framework that allows you do that has a nice, intuitive interpretation, that we will briefly outline in the following.

In this framework, we calculate the particle's probability amplitude $K(B,A)$ of going from the point A to the point B, term by term. First, although the particle has a potential energy, we assume that the particle's potential energy is zero, i.e., the particle is free. This is the 0_{th} order approximation. Hence, replacing the K's on the right of Eq. (1) with free particle propagators, we get

Feynman's Path Integral explained with basic Calculus

$\int_{-\infty}^{+\infty} dx_c K_0(B,C) K_0(C,A)$. (We use the subscript "0" here to indicate that the propagator corresponds to a free particle, not to be confused with how the subscript 0 is used to denote the action for the classical path or least path throughout the book.) Now, $\int_{-\infty}^{+\infty} dx_c K_0(B,C) K_0(C,A)$ is the $K_0(B,A)$, the free particle propagator from A to B. (The result is a consequence of using Eq. (1) for a free particle). So, in the 0_{th} order approximation, we can set $K(B,A)$ to the free particle propagator $K_0(B,A)$.

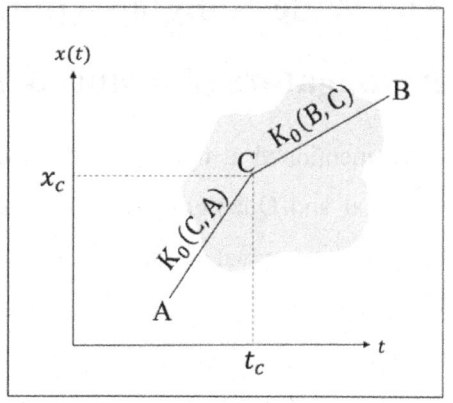

Figure 39

Next, we calculate the first order correction to $K_0(B,A)$ due to the potential in the following way: you assume that the particle goes freely from A to C (See Figure 39), momentarily experiences a potential, i.e., has a potential energy at C, then once again goes freely from C to B. As it turns out, the first order correction to the free-particle propagator $K_0(B,A)$ involves the product of:

1. the free particle propagator from A to C, i.e., $K_0(C,A)$
2. the particle's potential energy at C, i.e., $V(C)$ and
3. the free particle propagator from C to B, i.e., $K_0(B,C)$.

Finally, you need to sum up such contributions from all possible C's. The mathematical expression for the first order correction to the propagator is:

$$\frac{-i}{\hbar} \int_{t_a}^{t_b} dt_c \int_{-\infty}^{+\infty} dx_c\, K_0(B,C) V(C) K_0(C,A) \quad \text{......... Eq. (2)}.$$

Although Eq. (2) has some formal similarity with Eq. (1), there are clear differences: There is a potential energy and two integration variables (space and time) in Eq. (2) unlike in Eq. (1).

Appendix 4

The second order correction to the propagator $K(B,A)$ is calculated by considering the particle experiencing the potential twice: first at C and then at D (See Figure 40). Between C and D, the particle moves freely. To get the propagator for this process, you need to multiply the following together:

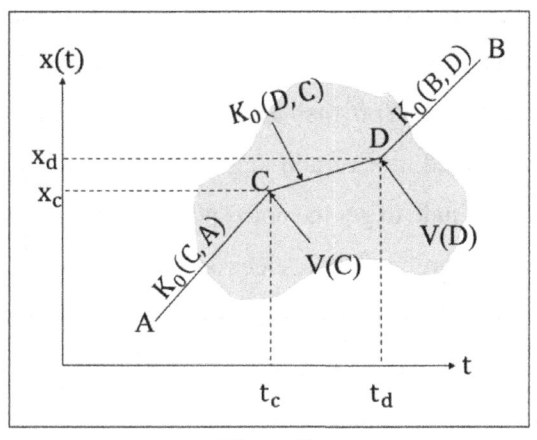

Figure 40

1. the free particle propagator from A to C, i.e., $K_0(C,A)$
2. the particle's potential energy at C ($V(C)$)
3. the free particle propagator from C to D, i.e., $K_0(D,C)$
4. the potential energy at D ($V(D)$), and
5. the free particle propagator from D to B, i.e., $K_0(B,D)$.

You need to sum over the contributions from the intermediate points C and D (integrate with respect to the corresponding space and time variables) with the constraint that the time corresponding to D (i.e., t_d) is greater than the time corresponding to C (i.e., t_c): $t_d > t_c$. The constraint arises from a non-relativistic particle's going only forward in time. (A relativistic particle has also a probability amplitude of going backward in time.)

So, you saw how free particle propagators continue to show up in the perturbative calculation of the propagator even when the particle isn't free, i.e., has a potential energy. In the following, we give a concrete "physical example" of that. Consider electrons hitting a thin metal foil. Subsequently, the positively charged cores of the metal-atoms attract the electrons, as a result of which, the electrons get scattered, i.e., change direction.

Feynman's Path Integral explained with basic Calculus

A quantity of interest for such problems is the probability of finding electrons at various angles with the original direction, after they have been scattered. You first calculate the electron's probability amplitude of being scattered at a certain angle, then take the absolute square of the probability amplitude to get the corresponding probability density.

Assuming that the electron beam is narrow, you can argue that the 0_{th} order term of the propagator will not contribute to the probability amplitude at a finite angle (shown in Figure 41). Because the beam is narrow, the 0_{th} order term of the propagator will be limited to the $\theta = 0$ region. Hence, when the electron is scattered at a significantly large θ, the contribution to the probability amplitude will come from the first order correction term of the propagator such as Eq. (2). To get the first order term for our problem, you need to multiply the following together.

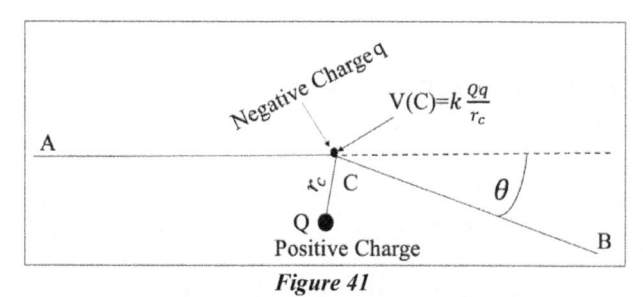

Figure 41

1. the free electron propagator from A to C ($K_0(C, A)$),
2. the electron's Coulombic potential energy $V(C) = k\frac{qQ}{r_c}$ at C (q and Q being the charges and r_c being the distance between them),
3. the free particle propagator from C to B ($K_0(B, C)$). The required probability amplitude is obtained from "one dimensional time plus three-dimensional spatial" version of Eq. (2) with the Coulomb potential energy for $V(C)$.

Appendix 5

Functional Derivatives

Note: Functional derivatives are not used in the book but are related to the book's content. Also, read Appendix 1 (on Functionals) before reading this Appendix.

How do you find the derivative of the function $f(x) = x^2$ (See Figure 42)? Here is one way. Changing x to a slightly different value $x + h$, (h being small), you get $f(x + h) = (x + h)^2 \approx x^2 + (2x)h$. [Keeping terms only up to the first order in h]. So, $f(x + h) \approx f(x) + (2x)h$, which implies that the change in f ($\equiv \Delta f = f(x + h) - f(x)$) is just $(2x)h$; the derivative of $f(x) \equiv \left(\frac{df}{dx}\right)$ is what multiplies h in the expression for Δf, i.e., $2x$.

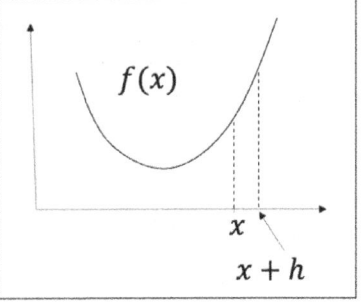

Figure 42

We will analogously introduce the derivative of functionals. Consider the functional $F[x(t)] = \int_{t_a}^{t_b} x^2(t)dt$, where $x(t)$ is defined over the interval (t_a, t_b) (See Figure 43). Let us change $x(t)$ to $x(t) + \eta(t)$, where $\eta(t)$ is a small

$\eta(t_a) = \eta(t_b) = 0$

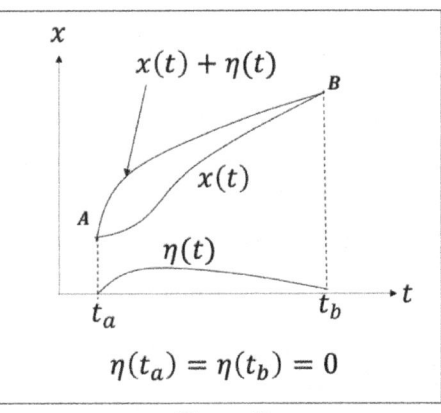

Figure 43

Feynman's Path Integral explained with basic Calculus

variation such that $\eta(t_a) = \eta(t_b) = 0$. Note how $\eta(t)$ is "analogous" to h in the previous example.

How much does $F[x(t)]$ change by as a result of changing $x(t)$ to $x(t) + \eta(t)$?

$$F[x(t) + \eta(t)]$$
$$= \int_{t_a}^{t_b}(x(t) + \eta(t))^2 dt = \int_{t_a}^{t_b}[x^2(t) + 2x(t)\eta(t) + \eta^2(t)]dt$$

Keeping only up to "linear-in-η" terms, i.e., ignoring the second order term $\eta^2(t)$, we get:

$$F[x(t) + \eta(t)] \approx \int_{t_a}^{t_b}[x^2(t)]dt + \int_{t_a}^{t_b} 2x(t)\eta(t)dt$$
$$= F[x(t)] + \int_{t_a}^{t_b} 2x(t)\eta(t)\, dt.$$

[We used $F[x(t)] = \int_{t_a}^{t_b} x^2(t)dt$]

.... Eq. (1).

From Eq. (1), the change in F, customarily written as $\delta F (\equiv F[x(t) + \eta(t)] - F[x(t)])$, is:

$$\delta F = \int_{t_a}^{t_b} 2x(t)\eta(t)\, dt \quad \ldots\ldots\ldots\ldots\ldots \text{Eq. (2).}$$

The functional derivative, notationally written as, $\frac{\delta F}{\delta x(t)}$, is defined as what multiplies $\eta(t)$ in the integrand of δF in Eq. (2) (note the analogy with the ordinary derivative: "$\left(\frac{df}{dx}\right)$ is what multiplies h in Δf"; $h \leftrightarrow \eta(t)$ and $\Delta f \leftrightarrow \delta F$). Hence, from Eq. (2), $\frac{\delta F}{\delta x(t)} = 2x(t)$.

Next, as another example, we consider the functional $F[x(t)] = \int_{t_a}^{t_b}\left(\frac{dx}{dt}\right)^2 dt$. Changing $x(t)$ to $x(t) + \eta(t)$,

Appendix 5

$$F[x(t) + \eta(t)] = \int_{t_a}^{t_b} \left(\frac{dx}{dt} + \frac{d\eta}{dt}\right)^2 dt$$

$$= \int_{t_a}^{t_b} \left[\left(\frac{dx}{dt}\right)^2 + 2\frac{dx}{dt}\frac{d\eta}{dt} + \left(\frac{d\eta}{dt}\right)^2\right] dt$$

Ignoring the "second order" term $\left(\frac{d\eta}{dt}\right)^2$ and keeping only up to the linear order term in η in $F[x(t) + \eta(t)]$, we get:

$$F[x(t) + \eta(t)] \approx \int_{t_a}^{t_b} \left(\frac{dx}{dt}\right)^2 dt + \int_{t_a}^{t_b} 2x(t)\eta(t) dt$$

$$= F[x(t)] + \int_{t_a}^{t_b} 2\frac{dx}{dt}\frac{d\eta}{dt} dt. \text{ [We used}$$

$$F[x(t)] = \int_{t_a}^{t_b} \left(\frac{dx}{dt}\right)^2 dt]$$

Hence, the change in F: $\delta F \equiv F[x(t) + \eta(t)] - F[x(t)] = \int_{t_a}^{t_b} 2\frac{dx}{dt}\frac{d\eta}{dt} dt$, which can be massaged into the following by using integration by parts: $\int_{t_a}^{t_b} \frac{dx}{dt}\frac{d\eta}{dt} dt = \frac{dx}{dt}\eta(t)\Big|_{t=t_a}^{t=t_b} - \int_{t_a}^{t_b} \eta(t)\frac{d}{dt}\left(\frac{dx}{dt}\right)$, the first term of which is zero, since $\eta(t_a) = \eta(t_b) = 0$. Hence, $\delta F = -2\int_{t_a}^{t_b} \eta(t)\frac{d}{dt}\left(\frac{dx}{dt}\right)$. Since $-2\frac{d^2x}{dt^2}$ multiplies $\eta(t)$ in the integral, per the definition of functional derivative, we have:

$$\frac{\delta F}{\delta x(t)} = -2\frac{d^2x}{dt^2}, \text{ when } F[x(t)] = \int_{t_a}^{t_b} \left(\frac{dx}{dt}\right)^2 dt \dots\dots\dots\dots\text{Eq. (3)}.$$

We know that derivatives of a function vanish at extrema. Along similar lines, a functional derivative vanishes for the function that minimizes a given functional. Hence, the functional derivative of the action functional should vanish for the least path or the classical path, since, per Least Action Principle (Chapter 2), the classical path minimizes the action functional. Hence, the vanishing functional derivative of the action functional should produce Newton's second law of motion leading to the classical path. We check that in the following by using a specific example.

Feynman's Path Integral explained with basic Calculus

Let us consider the action functional of a falling stone (Chapter 2), given by:

$$S[y(t)] = \int_{t_a}^{t_b} \left[\frac{1}{2}m\left(\frac{dy}{dt}\right)^2 + mgy\right]dt \quad \text{............... Eq. (4).}$$

To calculate the functional derivative of S, we evaluate S, as given by Eq. (4), for the path $(y(t) + \eta(t))$ that is slightly different from $y(t)$. $\eta(t)$, representing the small variation, satisfy $\eta(t_a) = \eta(t_b) = 0$. In the following, we only keep terms linear in $\eta(t)$ and its derivatives and ignore the higher order ones.

$$S[y(t) + \eta(t)]$$

$$= \int_{t_a}^{t_b} \left[\frac{1}{2}m\left(\frac{dy}{dt} + \frac{d\eta}{dt}\right)^2 + mg(y + \eta)\right]dt \text{ [We just replaced } y \text{ by}$$

$$y + \eta \text{ in Eq. (4)]}$$

$$= \int_{t_a}^{t_b} \left[\left(\frac{1}{2}m\left(\frac{dy}{dt}\right)^2 + mgy\right) + m\frac{dy}{dt}\frac{d\eta}{dt} + mg\eta\right]dt \text{ [We used the}$$

$$(a + b)^2 \text{ expansion formula,}$$

$$\text{ignored } \left(\frac{d\eta}{dt}\right)^2, \text{ then}$$

$$\text{regrouped terms.]}$$

$$= S[y(t)] + \int_{t_a}^{t_b} \left[m\frac{dy}{dt}\frac{d\eta}{dt} + mg\eta\right]dt$$

[We used Eq. (4) for $S[y(t)]$]

Hence, from the above, the change in S, given by δS, is:

$$\delta S \equiv S[y(t) + \eta(t)] - S[y(t)]$$

$$= \int_{t_a}^{t_b} \left[m\frac{dy}{dt}\frac{d\eta}{dt} + mg\eta\right]dt \quad \text{..................... Eq. (5).}$$

Integrating by parts, $\int_{t_a}^{t_b} \frac{dy}{dt}\frac{d\eta}{dt}dt = \frac{dy}{dt}\eta(t)\Big|_{t=t_a}^{t=t_b} - \int_{t_a}^{t_b} \eta(t)\frac{d}{dt}\left(\frac{dy}{dt}\right)$, the first term of which is zero, since $\eta(t_a) = \eta(t_b) = 0$. Hence, from Eq. (5),

$$\delta S = \int_{t_a}^{t_b} \eta(t)\left[-m\frac{d^2y}{dt^2} + mg\right]$$

Appendix 5

Since $\left(-m\frac{d^2y}{dt^2} + mg\right)$ multiplies $\eta(t)$ in the above integral, per the definition of functional derivative, we have $\frac{\delta S}{\delta y(t)} = -m\frac{d^2y}{dt^2} + mg$, setting which to zero, you get:

$$-m\frac{d^2y}{dt^2} + mg = 0, \text{ i.e.}$$

$$m\frac{d^2y}{dt^2} = mg \text{ (Newton's law in a constant gravitational field!)}$$

Similarly, calculating the functional derivative of the action for a spring-mass system, as given by, $S[x(t)] = \int_{t_a}^{t_b} \left[\frac{1}{2}m\left(\frac{dx}{dt}\right)^2 - \frac{1}{2}kx^2(t)\right]dt$, you should get:

$$\frac{\delta S}{\delta x(t)} = -m\frac{d^2x}{dt^2} - kx(t),$$ setting which to zero you will give you $m\frac{d^2x}{dt^2} = -kx(t)$ (Newton's law for a spring mass system.)

Feynman's Path Integral explained with basic Calculus

Appendix 6

Statistics of a Random Walk

This appendix assumes knowledge of Binomial probability distribution and concepts such as the expectation of a random variable on the reader's part. You can model the drunkard's random walk presented in Chapter 7 as follows: Suppose the drunkard took a total of N steps, of which some are to the right and some, to the left. The probability of taking a step to the right or to the left is the same, i.e., is equal to $\frac{1}{2}$. Suppose that out of the total N steps that the drunkard took, m are to the right. Hence, a total of $N - m$ steps were taken to the left. The problem, thus modelled, represents a Binomial distribution of N events with the "success probability" (the probability that a step is taken to the right) $p = \frac{1}{2}$.

The standard deviation σ of a Binomially distributed variable with the success probability p is $\sqrt{Np(1-p)}$. For the variable m in question, $p = \frac{1}{2}$, and hence the standard deviation of m, which is Binomially distributed, is

$$\sqrt{N \cdot \frac{1}{2} \cdot \frac{1}{2}} = \sqrt{\frac{N}{4}} = \frac{1}{2}\sqrt{N}.$$

The expectation or the average $E(m)$ of the Binomial variable m is $E(m) = Np = N\frac{1}{2}$.

Summarizing, the average and the standard deviation of m are $\frac{N}{2}$ and $\frac{1}{2}\sqrt{N}$ respectively. We will use them to find the probability distribution of the drunkard's distance from his initial position. Let the pole the drunkard starts from is located at $x = 0$; the drunkard's step-length is l. Since the drunkard

Feynman's Path Integral explained with basic Calculus

takes m out of a total of N steps to the right, he moves a distance of ml to the right and $(N - m)l$ to the left (See Figure 44). Hence, the net displacement of the drunkard from the pole is:

$$x = ml - (N - m)l = 2ml - Nl \dots \text{Eq. (1)}.$$

So, the average or the expectation (denoted by E) of the random variable $x = 2ml - Nl$ is:

$$E(x) = E[2lm - Nl]$$
$$= 2lE(m) - Nl \dots \text{Eq. (2)}.$$

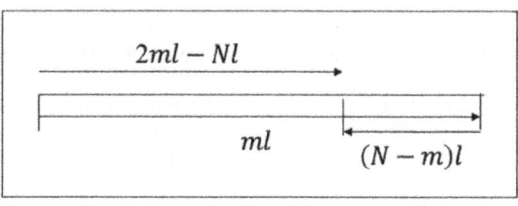

Figure 44

But as proved earlier, $E(m)$ (the average of m) is $N\frac{1}{2}$, using which in Eq. (2), we obtain:

$$E(x) = 2l \cdot \frac{N}{2} - Nl = 0 \dots \text{Eq. (3)}.$$

Eq. (3) proves that the average of the drunkard's position x is 0, i.e., where the drunkard started from. So, denoting the probability distribution for the drunkard's being at the location x, at time t as $P(x, t)$, the average or the mean of $P(x, t)$ is 0.

What about the spread of $P(x, t)$? We can calculate that using the variance of the random variable x or $Var(x)$ as follows:

$$Var(x) = E(x - E(x))^2 \text{ [Using the definition of variance]}$$
$$= E(2l(m - E(m)))^2 \text{ [Using Eq. (1) and Eq. (2)]}$$
$$= 4l^2 E(m - E(m))^2 \text{ [Bringing the constant in the front]}$$
$$= 4l^2 Var(m) \text{ [Using the definition of the variance of m]}$$
$$= 4l^2 (standard\ deviation\ of\ m)^2 \dots \text{Eq. (4)}.$$

In the last step, we used the fact that the variance of m is the square of its standard deviation, which we calculated as $\frac{1}{2}\sqrt{N}$ from Binomial distribution.

Appendix 6

Hence, from Eq. (4), we obtain $Var(x) = 4l^2 \left(\frac{1}{2}\sqrt{N}\right)^2 = l^2 N$, taking the square root of which, we obtain the standard deviation of x as follows.

$$standard\ deviation\ of\ x = l\sqrt{N} \dots\dots\dots\dots\dots\dots\dots\dots \text{Eq. (5)}.$$

We can assume that the total number of steps N increases linearly with the time t: $N \propto t$ (The longer the elapsed time, the more the number of steps the drunkard takes.) Hence, from Eq. (5), we obtain the standard deviation of x to be proportional to $\sqrt{t}$.

Hence, $P(x,t)$ has a mean or average of 0 and a standard deviation proportional to the square root of time elapsed: $\sqrt{t}$.

Feynman's Path Integral explained with basic Calculus

Appendix 7

Solutions

Chapter 1

Exercise 1: Evaluate $\int_{-\infty}^{+\infty} e^{-2x^2+6x} dx$ by directly completing the square, then using Eq. (1), i.e., $\int_{-\infty}^{+\infty} e^{-x^2} dx = \sqrt{\pi}$. Does the integral come out as (a constant)* $e^{4.5}$?

Soln. e^{-2x^2+6x}

$$= e^{-2\left[x^2 - 2\cdot\frac{3}{2}\cdot x + \left(\frac{3}{2}\right)^2 - \frac{9}{4}\right]}$$

$$= e^{-2\left[\left(x-\frac{3}{2}\right)^2 - \frac{9}{4}\right]}$$

$$= e^{-2\left[\left(x-\frac{3}{2}\right)^2\right]} e^{\frac{9}{2}} \quad \text{...................................... Eq. (1).}$$

Hence,

$$\int_{-\infty}^{+\infty} e^{-2x^2+6x} dx$$

$$= \int_{-\infty}^{+\infty} e^{-2\left[\left(x-\frac{3}{2}\right)^2\right]} e^{\frac{9}{2}} dx$$

$$= e^{\frac{9}{2}} \int_{-\infty}^{+\infty} e^{-2\left[\left(x-\frac{3}{2}\right)^2\right]} dx \quad \text{.. Eq. (2).}$$

Note that the factor $e^{\frac{9}{2}} = e^{4.5}$ has already shown up in the result. You can carry out the integral in Eq. (2) by substituting y for $\sqrt{2}\left(x - \frac{3}{2}\right)$. Then,

$$\int_{-\infty}^{+\infty} e^{-2\left[\left(x-\frac{3}{2}\right)^2\right]} dx = \frac{1}{\sqrt{2}} \int_{-\infty}^{+\infty} e^{-y^2} dy = \sqrt{\frac{\pi}{2}} \quad \text{........................ Eq. (3).}$$

Feynman's Path Integral explained with basic Calculus

We used $\int_{-\infty}^{+\infty} e^{-y^2} dy = \sqrt{\pi}$ in the last step. Using Eq. (3) in Eq. (2), we get

$\int_{-\infty}^{+\infty} e^{-2x^2+6x} dx = \sqrt{\frac{\pi}{2}} e^{4.5} = $ (a constant) $e^{4.5}$. So, yes, the integral does come out as (a constant) $e^{4.5}$.

Exercise 2: Prove by completing square that for $A > 0$, $\int_{-\infty}^{+\infty} e^{-Ax^2+Bx} dx = \sqrt{\frac{\pi}{A}} e^{\frac{B^2}{4A}}$

Soln. $-Ax^2 + Bx$

$$= -A\left(x^2 - 2.\frac{B}{2A} + \left(\frac{B}{2A}\right)^2 - \frac{B^2}{4A^2}\right)$$

$$= -A\left[\left(x - \frac{B}{2A}\right)^2 - \frac{B^2}{4A^2}\right]$$

$$= -A\left(x - \frac{B}{2A}\right)^2 + \frac{B^2}{4A} \quad \ldots \ldots \ldots \ldots \ldots \ldots \text{Eq. (4).}$$

Hence, using Eq. (4) in the following integral, we get:

$\int_{-\infty}^{+\infty} e^{-Ax^2+Bx} dx$

$= \int_{-\infty}^{+\infty} e^{-A\left(x-\frac{B}{2A}\right)^2 + \frac{B^2}{4A}} dx$

$= e^{\frac{B^2}{4A}} \int_{-\infty}^{+\infty} e^{-A\left(x-\frac{B}{2A}\right)^2} dx \ldots \ldots \ldots \ldots \ldots \ldots \text{Eq. (5).}$

To do the integral in Eq. (5), we substitute $y = \sqrt{A}\left(x - \frac{B}{2A}\right)$.

Hence, $\int_{-\infty}^{+\infty} e^{-A\left(x-\frac{B}{2A}\right)^2} dx = \frac{1}{\sqrt{A}} \int_{-\infty}^{+\infty} e^{-y^2} dy = \sqrt{\frac{\pi}{A}}$ (In the last step, we used $\int_{-\infty}^{+\infty} e^{-y^2} dy = \sqrt{\pi}$).

Using the above result in Eq. (5), we get $\int_{-\infty}^{+\infty} e^{-Ax^2+Bx} dx = e^{\frac{B^2}{4A}} \sqrt{\frac{\pi}{A}}$

(Proved).

Appendix 7

Exercise 3: Taking $x = 0.4$ calculate an approximate value for $\sqrt{1.4}$ using Eq. (7) of Chapter 1. (You can keep terms up to the third power of x) Then calculate $\sqrt{1.4}$ with your calculator and verify that the two results are close.

<u>Soln.</u> Using $x = .4$ and $n = \frac{1}{2}$ in the following series expansion (Eq. (7), Chapter 1)

$$(1+x)^n = 1 + nx + \frac{n(n-1)}{2!}x^2 + \frac{n(n-1)(n-2)}{3!}x^3 + \cdots,$$

we obtain:

$$(1+.4)^{\frac{1}{2}} \approx 1 + \frac{1}{2}(.4) + \frac{\frac{1}{2}\left(\frac{1}{2}-1\right)}{2!}(.4)^2 + \frac{\frac{1}{2}\left(\frac{1}{2}-1\right)\left(\frac{1}{2}-2\right)}{3!}(.4)^3 \text{ [Keeping terms up to } (0.4)^3]$$

$$= 1 + .2 - .02 + .004 = 1.184.$$

Using a calculator, $\sqrt{1.4} \approx 1.183$, which is close to what we got by the series expansion.

Exercise 4: Find the numerical coefficient for the x^5 term of $(1+x)^{\frac{1}{2}}$, using Eq. (7) (Chapter 1)

<u>Soln.</u> The coefficient of the x^5 term is:

$$\frac{\frac{1}{2}\left(\frac{1}{2}-1\right)\left(\frac{1}{2}-2\right)\left(\frac{1}{2}-3\right)\left(\frac{1}{2}-4\right)}{5!}x^5 = \frac{7}{256}x^5$$

Exercise 5. Expand $f(x) = (1+x)^{-1}$ about $x = 0$ using Eq. (9) in Chapter 1. If you think your Taylor expansion looks like an infinite GP series, you are right. What is the common ratio of the GP series you just found? For what values of x does the series converge? [Hint: the magnitude of the common ratio of an infinite GP series should be less than one for the series to converge.]

<u>Soln.</u> We will apply the Taylor expansion (Chapter 1 Eq. (9)) on $f(x) = (1+x)^{-1}$, the derivatives of which are:

Feynman's Path Integral explained with basic Calculus

$$\frac{df}{dx} = -(1+x)^{-2} \Rightarrow \frac{d^2f}{dx^2} = 2(1+x)^{-3} \Rightarrow \frac{d^3f}{dx^3} = 2(-3)(1+x)^{-4}$$

.....etc.

Evaluating $f(x)$ and its derivatives at $x = 0$, we get $f(x)|_{x=0} = 1$, $\frac{df}{dx}\big|_{x=0} = -1$, $\frac{d^2f}{dx^2}\big|_{x=0} = 2$, $\frac{d^3f}{dx^3}\big|_{x=0} = -6$, ...etc., using which in the Taylor series expansion, we get:

$$f(x) = f(x)|_{x=0} + \frac{df}{dx}\bigg|_{x=0} x + \frac{1}{2!}\frac{d^2f}{dx^2}\bigg|_{x=0} x^2 + \frac{1}{3!}\frac{d^3f}{dx^3}\bigg|_{x=0} x^3 + \cdots$$

$$= 1 - x + \frac{1}{2!}(2)x^2 + \frac{1}{3!}(-6)x^3 + \cdots \ldots$$

$$= 1 - x + x^2 - x^3 + \cdots \ldots$$

The above series expansion does look like an infinite GP series. As you can see from the expansion, the common ratio of the series is $-x$, the absolute value of which, i.e., $|x|$, should be less than one for the GP series to converge.

Exercise 6. Check by squaring each of $\frac{1}{\sqrt{2}} + i\frac{1}{\sqrt{2}}$ and $\frac{1}{\sqrt{2}} - i\frac{1}{\sqrt{2}}$ that you get i in each case.

Soln. $\left(\frac{1}{\sqrt{2}} + i\frac{1}{\sqrt{2}}\right)^2 = \left[\frac{1}{\sqrt{2}}(1+i)\right]^2 = \frac{1}{2}(1+i)^2 = \frac{1}{2}(1 + 2i + i^2) = \frac{1}{2} \cdot 2i = i$ [We used $(a+b)^2 = a^2 + 2ab + b^2$, and $i^2 = -1$]

Similarly, $\left(-\frac{1}{\sqrt{2}} - i\frac{1}{\sqrt{2}}\right)^2 = i.$

Exercise 7. Obtain first few terms of the Taylor expansion of $f(x) = \cos x$.

Soln. We will use the Taylor expansion (Eq. (11) of Chapter 1) with $a = 0$,

$f(x) = \cos x \Rightarrow \frac{df}{dx} = -\sin x \Rightarrow \frac{d^2f}{dx^2} = -\cos x \Rightarrow \frac{d^3f}{dx^3} = \sin x$ .....etc.

Evaluating $f(x) = \cos x$ and its derivatives at $x = 0$, we get $f(x)|_{x=0} = \cos 0 = 1$, $\frac{df}{dx}\big|_{x=0} = -\sin 0 = 0$, $\frac{d^2f}{dx^2}\big|_{x=0} = -\cos 0 = -1$, $\frac{d^3f}{dx^3}\big|_{x=0} = \sin 0 = 0$, ...etc. Using them in Eq. (11) of Chapter 1, we get:

Appendix 7

$$\cos x = 1 - \frac{1}{2!}x^2 + \cdots$$

Additional Exercises

1. Prove $\int_{-\infty}^{+\infty} dx e^{[-k_1(x-a)^2 - k_2(x-b)^2]} = \sqrt{\frac{\pi}{k_1+k_2}} e^{\frac{-k_1 k_2}{k_1+k_2}(a-b)^2}$

<u>Soln.</u> We expand $-k_1(x-a)^2 - k_2(x-b)^2$ as

$$-k_1[x^2 - 2ax + a^2] - k_2[x^2 - 2bx + b^2]$$
$$= -k_1[x^2 - 2ax + a^2] - k_2[x^2 - 2bx + b^2]$$
$$= -x^2(k_1 + k_2) + 2x(ak_1 + bk_2) - (a^2 k_1 + b^2 k_2)$$
$$\equiv -Ax^2 + Bx + C \text{, where,}$$

$A = (k_1 + k_2), B = 2(ak_1 + bk_2)$ and $C = -(a^2 k_1 + b^2 k_2)$

.... Eq. (6).

Hence,

$$\int_{-\infty}^{+\infty} dx e^{[-k_1(x-a)^2 - k_2(x-b)^2]} = \int_{-\infty}^{+\infty} dx e^{-Ax^2+Bx+C} = e^C \int_{-\infty}^{+\infty} dx e^{-Ax^2+Bx}$$

.... Eq. (7).

Evaluating Eq. (7) using $\int_{-\infty}^{+\infty} dx e^{-Ax^2+Bx} = \sqrt{\frac{\pi}{A}} e^{\frac{B^2}{4A}}$ (Eq. (6), Chapter 1),

we get:

$$\int_{-\infty}^{+\infty} dx e^{-Ax^2+Bx+C} = \sqrt{\frac{\pi}{A}} e^{\frac{B^2}{4A}} e^C = \sqrt{\frac{\pi}{A}} e^{C + \frac{B^2}{4A}} \ldots\ldots\ldots\ldots \text{ Eq. (8).}$$

Finally, substituting back the expressions for A, B and C, given by Eq. (6) (of the current Appendix), in Eq. (8), we get:

$$\int_{-\infty}^{+\infty} dx e^{[-k_1(x-a)^2 - k_2(x-b)^2]} = \sqrt{\frac{\pi}{k_1+k_2}} e^{-(a^2 k_1 + b^2 k_2) + \frac{(ak_1+bk_2)^2}{k_1+k_2}}$$

$$= \sqrt{\frac{\pi}{k_1+k_2}} e^{-\frac{k_1 k_2}{k_1+k_2}(a-b)^2} \quad \text{(Proved)}$$

2. Integrate $\int_{-\infty}^{+\infty} \int_{-\infty}^{+\infty} dx dy e^{i[(x-a)^2 + (x-y)^2 + (y-b)^2]}$

Feynman's Path Integral explained with basic Calculus

<u>Soln.</u> We write the above double integral as two single integrals as follows:

$\int_{-\infty}^{+\infty}\int_{-\infty}^{+\infty} dxdy\, e^{i[(x-a)^2+(x-y)^2+(y-b)^2]}$

$= \int_{-\infty}^{+\infty} dx\, e^{i(x-a)^2} \int_{-\infty}^{+\infty} dy\, e^{i[(x-y)^2+(y-b)^2]}$ .... Eq. (9).

Using Eq. (17) of Chapter 1, viz., $\int_{-\infty}^{+\infty} du\, e^{i[p_1(u-a)^2+p_2(u-b)^2]} = \sqrt{\dfrac{i\pi}{p_1+p_2}}\, e^{\frac{ip_1 p_2}{p_1+p_2}(a-b)^2}$, with $p_1 = p_2 = 1$, we carry out the y-integration

(treating x as constant) in Eq. (9) to obtain $\int_{-\infty}^{+\infty} dy\, e^{i[(x-y)^2+(y-b)^2]} = \sqrt{\dfrac{i\pi}{2}}\, e^{\frac{i}{2}(x-b)^2}$. Hence, the integral w.r.t x, which we should carry out next, is given by the following:

$\sqrt{\dfrac{i\pi}{2}} \int_{-\infty}^{+\infty} dx\, e^{i(x-a)^2}\, e^{\frac{i}{2}(x-b)^2}$ .... Eq. (10).

Once again, we use Eq. (17) of Chapter 1, but this time with $p_1 = 1$, and $p_2 = \dfrac{1}{2}$, to evaluate Eq. (10) (of the current Appendix) as:

$\sqrt{\dfrac{i\pi}{2}} \int_{-\infty}^{+\infty} dx\, e^{i(x-a)^2}\, e^{\frac{i}{2}(x-b)^2} = \sqrt{\dfrac{i\pi}{2}}\sqrt{\dfrac{2i\pi}{3}}\, e^{\frac{i}{3}(a-b)^2} = \dfrac{i\pi}{\sqrt{3}}\, e^{\frac{i}{3}(a-b)^2}$

(Ans.)

Chapter 2

Exercise 1. Using $\bar{y}(x = x_a) = y_a$ and $\bar{y}(x = x_b) = y_b$ in $\bar{y}(x) = Kx + Q$, find out K and Q.

<u>Soln.</u> We have, $\bar{y}(x = x_a) = y_a = Kx_a + Q$, and $\bar{y}(x = x_b) = y_b = Kx_b + Q$. So, we have the following system of equations to solve.

$$y_a = Kx_a + Q$$
$$y_b = Kx_b + Q$$

Solving for K and Q, we get $K = \dfrac{y_b - y_a}{x_b - x_a}$, and $Q = \dfrac{y_a x_b - y_b x_a}{x_b - x_a}$

Appendix 7

Exercise 2. The solution to Eq. (22), Chapter 2 is given by $\bar{y}(t) = Kt + Q$. Find K and Q, given that $\bar{y}(t = t_a) = y_a$ and $\bar{y}(t = t_b) = y_b$. Then evaluate the integral S_0. i.e., calculate: $\int_{t=t_a}^{t=t_b} \left[\frac{1}{2} m \left(\frac{d\bar{y}}{dt}\right)^2\right] dt$

<u>Soln.</u> We have, $\bar{y}(t = t_a) = y_a = Kt_a + Q$, and $\bar{y}(t = t_b) = y_b = Kt_b + Q$. So, we have the following system of equations to solve.

$$y_a = Kt_a + Q$$
$$y_b = Kt_b + Q$$

Solving for K and Q, we get $K = \frac{y_b - y_a}{t_b - t_a}$, and $Q = \frac{y_a t_b - y_b t_a}{t_b - t_a}$

Now, since $\bar{y}(t) = Kt + Q$, $\frac{d\bar{y}}{dt} = K$, which was just calculated to be $\frac{y_b - y_a}{t_b - t_a}$. Replacing $\frac{d\bar{y}}{dt}$ by $\frac{y_b - y_a}{t_b - t_a}$ in $S_0 = \int_{t=t_a}^{t=t_b} \left[\frac{1}{2} m \left(\frac{d\bar{y}}{dt}\right)^2\right] dt$, we get: $S_0 = \frac{1}{2} m \frac{(y_b - y_a)^2}{t_b - t_a}$.

Exercise 3. Find C_1 and C_2 in the equation $\bar{y}(t) = \frac{1}{2} gt^2 + C_1 t + C_2$, given that $\bar{y}(t = t_a) = y_a$ and $\bar{y}(t = t_b) = y_b$. Then evaluate $S_0 = \int_{t=t_a}^{t=t_b} \left[\frac{1}{2} m \left(\frac{d\bar{y}}{dt}\right)^2 + mg\bar{y}\right] dt$.

<u>Soln.</u> Setting $\bar{y}(t = t_a) = \frac{1}{2} gt_a^2 + C_1 t_a + C_2 = y_a$, and

$$\bar{y}(t = t_b) = \frac{1}{2} gt_b^2 + C_1 t_b + C_2 = y_b,$$

we have the following system of equations:

$$C_1 t_a + C_2 = y_a - \frac{1}{2} gt_a^2, \text{ and}$$

$$C_1 t_b + C_2 = y_b - \frac{1}{2} gt_b^2 \text{ , solving which for } C_1 \text{ and } C_2 \text{, we get}$$

$$C_1 = \frac{y_b - y_a}{t_b - t_a} - \frac{1}{2} g(t_a + t_b), \text{ and } C_2 = \frac{y_a t_b - y_b t_a}{t_b - t_a} + \frac{1}{2} g t_b t_a$$

.... Eq. (1).

Next, using $\bar{y}(t) = \frac{1}{2} gt^2 + C_1 t + C_2$ and $\frac{d\bar{y}}{dt} = gt + C_1$ in S_0, we get:

Feynman's Path Integral explained with basic Calculus

$$S_0 = \int_{t=t_a}^{t=t_b} \left[\frac{1}{2}m\left(\frac{d\bar{y}}{dt}\right)^2 + mg\bar{y}\right]dt$$

$$= \frac{mg^2}{3}(t_b^3 - t_a^3) + mg(t_b^2 - t_a^2)C_1 + \left(\frac{1}{2}mC_1^2 + mgC_2\right)(t_b - t_a),$$

where C_1 and C_2 are given by Eq. (1)

Exercise 4. Check that $\bar{x} = A \cos \omega t + B \sin \omega t$, with $\omega = \sqrt{\frac{k}{m}}$, satisfies the differential equation $m\frac{d^2\bar{x}}{dt^2} + k\bar{x} = 0$ Find A and B given that $\bar{x}(t=0) = 0$ and $\bar{x}(t=t_b) = x_b$. Then evaluate $S_0 = \int_{t=t_a}^{t=t_b}\left[\frac{1}{2}m\left(\frac{d\bar{x}}{dt}\right)^2 - \frac{1}{2}k\bar{x}^2\right]dt$.

<u>Soln.</u> Since, $\bar{x} = A \cos \omega t + B \sin \omega t$,

$$\frac{d\bar{x}}{dt} = -A\omega \sin \omega t + B\omega \cos \omega t, \text{ which implies:}$$

$$\frac{d^2\bar{x}}{dt^2} = -A\omega^2 \cos \omega t - B\omega^2 \sin \omega t$$

$$= -\omega^2(A \cos \omega t + B \sin \omega t)$$

$$= -\frac{k}{m}\bar{x} \text{ [Using } \bar{x} = A \cos \omega t + B \sin \omega t \text{ and } \omega = \sqrt{\frac{k}{m}}]$$

Hence, we have $\frac{d^2\bar{x}}{dt^2} = -\frac{k}{m}\bar{x}$, from which follows: $m\frac{d^2\bar{x}}{dt^2} + k\bar{x} = 0$ (Proved) Next we will use $\bar{x}(t=0) = 0$ and $\bar{x}(t=t_b) = x_b$ in $\bar{x} = A \cos \omega t + B \sin \omega t$ to ascertain the constants A and B.

$\bar{x}(t=0) = (A \cos \omega t + B \sin \omega t)|_{t=0} = A = 0$. Hence, $\bar{x} = B \sin \omega t$, using $\bar{x}(t=t_b) = x_b$ in which, we get $x_b = B \sin \omega t_b$. Solving for B, we get $B = \frac{x_b}{\sin \omega t_b}$, using which in $\bar{x} = B \sin \omega t$, we get:

$$\bar{x} = \frac{x_b \sin \omega t}{\sin \omega t_b} \quad \ldots\ldots\ldots\ldots\ldots\ldots\ldots\ldots\ldots\ldots \text{ Eq. (2)}.$$

From Eq. (2), we obtain $\frac{d\bar{x}}{dt} = \frac{x_b \omega \cos \omega t}{\sin \omega t_b}$ $\ldots\ldots\ldots\ldots\ldots\ldots\ldots\ldots$ Eq. (3).

Appendix 7

Using Eq. (2) and Eq. (3) in $S_0 = \int_{t=0}^{t=t_b} \left[\frac{1}{2}m\left(\frac{dx}{dt}\right)^2 - \frac{1}{2}k\bar{x}^2\right]dt$, and

replacing k by $m\omega^2$ (since $\omega = \sqrt{\frac{k}{m}}$), we get:

$$S_0 = \frac{m\omega^2 x_b^2}{2(\sin \omega t_b)^2} \int_{t=0}^{t=t_b} [\cos^2 \omega t - \sin^2 \omega t] dt$$

Using the trig. identity $\cos^2 \omega t - \sin^2 \omega t = \cos 2\omega t$, and then carrying out the above integral, we get:

$$S_0 = \frac{m\omega x_b^2 \sin 2\omega t_b}{4(\sin \omega t_b)^2}$$

$$= \frac{m\omega x_b^2 \cos \omega t_b}{2 \sin \omega t_b} \quad \text{[In the last step, we used the trig. Identity:}$$

$\sin 2\omega t_b = 2 \sin \omega t_b \cos \omega t_b.$]

Chapter 3

Exercise 1. Assume that the time interval $t_b - t_a$ is divided into four equal parts ($N = 4$), each of length ϵ. The endpoints of the time-interval are $t_0 = t_a$ and $t_4 = t_b$. The intermediate time-points are $t_1, t_2,$ and t_3. $x(t_0) = x_a$ and $x(t_4) = x_b$. The intermediate x-values at time-points $t_1, t_2,$ and t_3 are denoted as $x_1, x_2,$ and x_3 respectively.

The action can be written follows.

$$S = \left(\frac{1}{2}m\right)\left[\left(\frac{x_1-x_a}{\epsilon}\right)^2 + \left(\frac{x_2-x_1}{\epsilon}\right)^2 + \left(\frac{x_3-x_2}{\epsilon}\right)^2 + \left(\frac{x_b-x_3}{\epsilon}\right)^2\right]\epsilon$$

$$= \left(\frac{1}{2}m\right)\left[\frac{(x_1-x_a)^2}{\epsilon} + \frac{(x_2-x_1)^2}{\epsilon} + \frac{(x_3-x_2)^2}{\epsilon} + \frac{(x_b-x_3)^2}{\epsilon}\right]$$

Carry out the following iterated multiple integral:

$$I = \left(\frac{1}{A}\right)^4 \int_{-\infty}^{+\infty} dx_1 \int_{-\infty}^{+\infty} dx_2 \int_{-\infty}^{+\infty} dx_3 \, e^{-kS}$$

$$I = \left(\frac{1}{A}\right)^4 \int_{-\infty}^{+\infty} dx_1 \int_{-\infty}^{+\infty} dx_2 \int_{-\infty}^{+\infty} dx_3$$

Feynman's Path Integral explained with basic Calculus

$$e^{-\frac{km}{2}\left[\frac{(x_1-x_a)^2}{\epsilon}+\frac{(x_2-x_1)^2}{\epsilon}+\frac{(x_3-x_2)^2}{\epsilon}+\frac{(x_b-x_3)^2}{\epsilon}\right]},$$

where $\frac{1}{A} \equiv \sqrt{\frac{km}{2\pi\epsilon}}$ (Eq. (6), Chapter 3). You will need to repeatedly use Eq. (4) from Chapter 3, i.e. $\int_{-\infty}^{+\infty} dx e^{[-k_1(x-a)^2-k_2(x-b)^2]} = \sqrt{\frac{\pi}{k_1+k_2}} e^{\frac{-k_1 k_2}{k_1+k_2}(a-b)^2}$.

Check that your final result is formally similar to Eq. (15) of Chapter 3.

<u>Soln.</u> We will integrate w.r.t x_3 first, treating the other variables x_1 and x_2 as constants. We write the integral of the problem as:

$I =$

$\left(\frac{1}{A}\right)^4 \int_{-\infty}^{+\infty} dx_1 dx_2 e^{-\left(\frac{km}{2\epsilon}\right)[(x_1-x_a)^2+(x_2-x_1)^2]} \int_{-\infty}^{+\infty} dx_3 \, e^{-\left(\frac{km}{2\epsilon}\right)[(x_3-x_2)^2+(x_b-x_3)^2]}$

.... Eq. (1).

We will carry out the x_3 integral in Eq. (1) with the help of the following:

$\int_{-\infty}^{+\infty} dx e^{[-k_1(x-a)^2-k_2(x-b)^2]} = \sqrt{\frac{\pi}{k_1+k_2}} e^{\frac{-k_1 k_2}{k_1+k_2}(a-b)^2}$ .... Eq. (2).

Setting $k_1 = k_2 = \frac{km}{2\epsilon}$ in Eq. (2), we obtain:

$\int_{-\infty}^{+\infty} dx_3 \, e^{-\left(\frac{km}{2\epsilon}\right)[(x_3-x_2)^2+(x_b-x_3)^2]} = \sqrt{\frac{\pi\epsilon}{km}} e^{-\frac{km}{2(2\epsilon)}(x_b-x_2)^2}$, using which in Eq. (1), we get

I

$= \left(\frac{1}{A}\right)^4 \sqrt{\frac{\pi\epsilon}{km}} \int_{-\infty}^{+\infty} dx_1 dx_2 e^{-\left(\frac{km}{2\epsilon}\right)[(x_1-x_a)^2+(x_2-x_1)^2]} e^{-\frac{km}{2(2\epsilon)}(x_b-x_2)^2}$

$= \left(\frac{1}{A}\right)^4 \sqrt{\frac{\pi\epsilon}{km}} \int_{-\infty}^{+\infty} dx_1 e^{-\left(\frac{km}{2\epsilon}\right)[(x_1-x_a)^2]} \int_{-\infty}^{+\infty} dx_2 e^{-\left(\frac{km}{2\epsilon}\right)(x_2-x_1)^2} e^{-\frac{km}{2(2\epsilon)}(x_b-x_2)^2}$

.... Eq. (3).

We will do the x_2 integral in Eq. (3) by using Eq. (2) with $k_1 = \frac{km}{2\epsilon}$, and $k_2 = \frac{km}{2(2\epsilon)}$. We get:

Appendix 7

$$\int_{-\infty}^{+\infty} dx_2 \, e^{-\left(\frac{km}{2\epsilon}\right)(x_2-x_1)^2} \, e^{-\frac{km}{2(2\epsilon)}(x_b-x_2)^2} = \sqrt{\frac{4\pi\epsilon}{3km}} \, e^{-\frac{km}{2(3\epsilon)}(x_b-x_1)^2}, \text{ using which}$$

in Eq. (3), we get:

$$I = \left(\frac{1}{A}\right)^4 \sqrt{\frac{\pi\epsilon}{km}} \sqrt{\frac{4\pi\epsilon}{3km}} \int_{-\infty}^{+\infty} dx_1 \, e^{-\left(\frac{km}{2\epsilon}\right)[(x_1-x_a)^2]} \, e^{-\frac{km}{2(3\epsilon)}(x_b-x_1)^2} \quad \ldots \ldots \text{ Eq. (4).}$$

We carry out the x_1 integral in Eq. (4) by setting $k_1 = \frac{km}{2\epsilon}$, and $k_2 = \frac{km}{2(3\epsilon)}$ in Eq. (2). We get:

$$I = \left(\frac{1}{A}\right)^4 \sqrt{\frac{\pi\epsilon}{km}} \sqrt{\frac{4\pi\epsilon}{3km}} \int_{-\infty}^{+\infty} dx_1 \, e^{-\left(\frac{km}{2\epsilon}\right)[(x_1-x_a)^2]} \, e^{-\frac{km}{2(3\epsilon)}(x_b-x_1)^2}$$

$$= \left(\frac{1}{A}\right)^4 \sqrt{\frac{\pi\epsilon}{km}} \sqrt{\frac{4\pi\epsilon}{3km}} \sqrt{\frac{3\pi\epsilon}{2km}} \, e^{-\frac{km}{2(4\epsilon)}(x_b-x_a)^2}$$

Using $\frac{1}{A} \equiv \sqrt{\frac{km}{2\pi\epsilon}}$ (Eq. (6) of Chapter 3) in the above, we get $I = \sqrt{\frac{km}{2\pi(4\epsilon)}} \, e^{-\frac{km}{2(4\epsilon)}(x_b-x_a)^2}$. But since, $4\epsilon = t_b - t_a$, we have $I = \sqrt{\frac{km}{2\pi(t_b-t_a)}} \, e^{-\frac{km}{2(t_b-t_a)}(x_b-x_a)^2}$, which is just what we have on the right side of Eq. (15) of Chapter 3.

Chapter 7

Exercise 1. Carry out the integral in Eq. (6) of Chapter 7, and derive Eq. (7)

Soln. From Eq. (6) of Chapter 7,

$$P_{neg} = \frac{1}{\sigma\sqrt{2\pi}} \int_{-\infty}^{+\infty} dV \, e^{-\left[CV + \frac{(V-V_0)^2}{2\sigma^2}\right]} \quad \ldots \ldots \ldots \text{ Eq. (1).}$$

By expanding $(V - V_0)^2$ in the argument of the exponential in Eq. (1), you get the following after a little algebra:

$$P_{neg} = \frac{1}{\sigma\sqrt{2\pi}} e^{-\frac{V_0^2}{2\sigma^2}} \int_{-\infty}^{+\infty} dV \, e^{-AV^2 + BV} \quad \ldots \ldots \ldots \text{ Eq. (2),}$$

where $A = \frac{1}{2\sigma^2}$, and $B = -\left(C - \frac{V_0}{\sigma^2}\right)$ .... Eq. (3).

Feynman's Path Integral explained with basic Calculus

Using $\int_{-\infty}^{+\infty} e^{-AV^2+BV} = \sqrt{\frac{\pi}{A}} e^{\frac{B^2}{4A}}$ (Eq. (6) from Chapter 1), with A and B given by Eq. (3), in Eq. (2), we get:

$$P_{neg} = e^{-\frac{V_0^2}{2\sigma^2}} e^{\frac{\sigma^2}{2}\left(C-\frac{V_0}{\sigma^2}\right)^2}, \text{ simplifying which, we get:}$$

$$P_{neg} = e^{\left(\frac{1}{2}\sigma^2 C^2 - CV_0\right)} \text{ (Proved.)}$$

Exercise 2. Check Eq. (14) from Chapter 7 is correct. Use the following formula you encountered in Eq. (16) of Chapter 1:

$$\int_{-\infty}^{+\infty} dx e^{[-k_1(x-a)^2 - k_2(x-b)^2]} = \sqrt{\frac{\pi}{k_1+k_2}} e^{\frac{-k_1 k_2}{k_1+k_2}(a-b)^2}$$

Soln. $P(t_b, x_b; t_c, x_c) = \dfrac{1}{D\sqrt{t_b-t_c}\sqrt{2\pi}} e^{-\frac{1}{2D^2}\frac{(x_b-x_c)^2}{(t_b-t_c)}}$

$P(t_c, x_c; t_a, x_a) = \dfrac{1}{D\sqrt{t_c-t_a}\sqrt{2\pi}} e^{-\frac{1}{2D^2}\frac{(x_c-x_a)^2}{(t_c-t_a)}}$

Using the above expressions,

$\int_{-\infty}^{+\infty} dx_c P(t_b, x_b; t_c, x_c) P(t_c, x_c; t_a, x_a)$

$= \dfrac{1}{D^2 2\pi \sqrt{t_b-t_c}\sqrt{t_c-t_a}} \int_{-\infty}^{+\infty} dx_c \, e^{-\frac{1}{2D^2}\frac{(x_b-x_c)^2}{(t_b-t_c)}} e^{-\frac{1}{2D^2}\frac{(x_c-x_a)^2}{(t_c-t_a)}} \dots\dots\dots$ Eq. (4).

Using the integral formula given at the beginning of this problem with $k_1 = \dfrac{1}{2D^2}\dfrac{1}{(t_b-t_c)}$ and $k_2 = \dfrac{1}{2D^2}\dfrac{1}{(t_c-t_a)}$, we get:

$$\int_{-\infty}^{+\infty} dx_c \, e^{-\frac{1}{2D^2}\frac{(x_b-x_c)^2}{(t_b-t_c)}} e^{-\frac{1}{2D^2}\frac{(x_c-x_a)^2}{(t_c-t_a)}}$$

$$= \sqrt{D^2 2\pi} \sqrt{\frac{(t_b-t_c)(t_c-t_a)}{t_b-t_a}} e^{-\frac{(x_b-x_a)^2}{2D^2(t_b-t_a)}} \dots\dots\dots\text{ Eq. (5).}$$

Using Eq. (5) in Eq. (4), we obtain:

Appendix 7

$$\int_{-\infty}^{+\infty} dx_c P(t_b, x_b; t_c, x_c) P(t_c, x_c; t_a, x_a) = \frac{1}{\sqrt{D^2 2\pi}} \sqrt{\frac{1}{t_b - t_a}} e^{-\frac{(x_b - x_a)^2}{2D^2(t_b - t_a)}} =$$

$\frac{1}{D\sqrt{2\pi}} \frac{1}{\sqrt{t_b - t_a}} e^{-\frac{(x_b - x_a)^2}{2D^2(t_b - t_a)}}$, which is precisely the expression for $P(t_b, x_b; t_a, x_a)$ (Proved).

Index

approximate, 8, 128, 140, 155
average, 120, 122, 123, 124, 151, 152
Binomial, 6, 7, 8, 9, 122, 151
Brownian motion, 131
classical, 3, 4, 20, 37, 64, 65, 66, 76, 77, 118, 138, 148
complete square, 4
completing square, 5, 6, 154, 155
complex exponential functions, 70
complex numbers, 3, 4, 17, 65, 66
Complex numbers, 11, 14
conditional probability, 118, 124, 130
constant, 3, 5, 14, 15, 17, 18, 20, 24, 29, 30, 33, 36, 38, 47, 50, 54, 55, 56, 58, 61, 65, 75, 76, 78, 79, 93, 94, 101, 103, 108, 109, 110, 113, 114, 115, 124, 126, 128, 135, 140, 150, 154, 159
convergence, 6, 16
De Broglie's, 74
definitive, 64
derivatives, 8, 9, 10, 12, 23, 27, 31, 33, 39, 104, 109, 114, 146, 148, 149, 156, 157
discretize, 44, 49, 53, 70, 81
discretize the action, 44, 53
electric field, 78, 134
energy, 4, 36, 67, 74, 77, 78, 79, 81, 89, 110, 115, 116, 134, 135, 142
Euler's identity, 13, 14
exponentials, 62, 65
falling stone, 78, 149

first order, 20, 21, 22, 24, 25, 27, 31, 59, 67, 92, 97, 105, 143, 145, 146
free particle, 4, 37, 38, 44, 45, 55, 67, 68, 69, 71, 72, 73, 74, 78, 81, 83, 89, 90, 101, 103, 105, 106, 107, 109, 112, 114, 116, 129, 137, 138, 142, 143, 144, 145
free-particle action, 38, 47, 49, 56, 71
functional, 21, 22, 24, 132, 133, 146, 147, 148, 149, 150
Functional Derivatives, 146
Gaussian, 4, 2, 3, 5, 14, 15, 47, 48, 56, 69, 119, 124, 125, 128, 129, 140
indirect method, 56, 61, 83, 90, 91, 95, 96
infinitesimal propagators, 72, 83
infnitesimal, 72
intermediate position variables, 73
jagged path, 45, 47, 49, 73
kinetic energy, 4, 36, 38, 71, 77, 78, 81, 90
Least Action Principle, 37, 38, 148
least path, 25, 26, 27, 30, 31, 32, 33, 34, 35, 37, 38, 39, 57, 58, 59, 65, 66, 91, 92, 93, 95, 96, 97, 148
linear term, 59, 93
line-segments, 49, 65, 70, 71, 72, 82, 83
maximum, 5, 20
mid-point, 81
minimizes, 22, 24, 26, 30, 32, 33, 36, 37, 65, 148
minimum, 20, 21, 24, 25, 26, 30, 33, 39, 60, 91
multiple variables, 53

Feynman's Path Integral explained with basic Calculus

negatives, 120, 122, 124
Newton's law, 3, 36, 37, 38, 64, 134, 150
normalization factor, 72, 83
normalized, 3, 48, 124
parabola, 35, 59
partial derivative, 101
Perturbation theory, 142
Poisson Plus, 123, 124, 126
potential energy, 4, 36, 38, 67, 71, 78, 79, 80, 81, 82, 83, 86, 89, 90, 101, 106, 107, 110, 111, 112, 116, 134, 135, 142, 143, 144, 145
probability amplitudes, 64, 89, 119, 138
probability density function, 74
propagator, 4, 5, 64, 65, 67, 72, 73, 74, 82, 83, 88, 89, 90, 91, 95, 96, 100, 101, 103, 106, 109, 110, 118, 119, 129, 138, 143, 144, 145
quadratic potential energy, 79
quantized, 115
Quantum mechanics, 4, 5, 62, 64, 89, 136
randomly, 121, 128
scattered, 144, 145

Schrodinger's equation, 5, 18, 67, 83, 100, 103, 105, 106, 110, 112, 115, 116
second-order, 147
series, 7, 8, 9, 102, 155, 156, 157
shortest, 20, 21, 24, 26, 29
spring mass, 135, 150
square, 3, 5, 11, 15, 23, 59, 64, 74, 75, 90, 93, 100, 126, 127, 128, 138, 140, 145, 153
stationarity, 24, 25, 26, 27, 28, 30, 31, 32, 33, 34, 39, 59, 66, 92, 93, 97
stationary, 20, 24, 25, 34, 40
Taylor, 3, 8, 9, 12, 13, 23, 27, 77, 101, 103, 106, 108, 110, 111, 113, 126, 156, 157
trajectories, 64
uncertainty, 136, 137, 138, 140, 141
variation, 20, 22, 24, 26, 27, 31, 33, 39, 59, 91, 92, 93, 96, 97, 147, 149
Variational Calculus, 4, 5, 20, 25, 30, 37
variations, 27, 31, 34, 66
wavefunction, 100, 101, 118
wavelength, 3, 4, 74, 75, 76

About the Author

Swapnonil Banerjee has a Ph.D. in Physics from the University of California, Davis, and many years of teaching experience from K12 to university physics courses. He is passionate about simplifying complex ideas, which served as a primary motivation for writing this book.

Swapnonil has extensive research experience in theoretical physics. He made important contributions to the study of a new material called semi-Dirac, in which electrons behave as non-relativistic, massive particles or effectively relativistic, massless particles depending on the direction of their movement. In one direction, the electron's energy-momentum relationship is indicative of the electron's having a definitive mass; in the orthogonal direction, the electron's energy-momentum relationship is indicative of a surprising absence of mass.

Swapnonil has also contributed to biotechnology, co-developing the Poisson Plus algorithm used for estimating the concentration of biomolecules via the digital PCR technique and holds a patent for this work.

Swapnonil has published in reputed journals including Nature Scientific Reports and Physical Review Letters.

Swapnonil's interests besides research and teaching include people and culture. He co-authored the historical fiction *Deflected*, a fast-paced, wartime romance based on the life of the nineteenth-century mathematician who calculated the height of Mount Everest and established it as the highest point on Earth.

Email swapno.banerjee@gmail.com

Connect: https://www.linkedin.com/in/swapnonil-banerjee-phd-5597553b/

Made in the USA
Las Vegas, NV
30 December 2024